Human Body Systems

Structure, Function and Environment

Daniel D. Chiras

University of Denver

JONES AND BARTLETT PUBLISHERS

Sudbury, Massachusetts

BOSTON TORONTO LONDON SINGAPORE

World Headquarters
Jones and Bartlett Publishers
40 Tall Pine Drive
Sudbury, MA 01776
978-443-5000
info@jbpub.com
www.jbpub.com

Jones and Bartlett Publishers Canada
2406 Nikanna Road
Mississauga, ON L5C 2W6
CANADA

Jones and Bartlett Publishers International
Barb House, Barb Mews
London W6 7PA
UK

Production Credits
Chief Executive Officer: Clayton Jones
Chief Operating Officer: Don W. Jones, Jr.
Executive V.P. & Publisher: Robert W. Holland, Jr.
Executive Editor, Science: Stephen L. Weaver
Associate Managing Editor, College Editorial: Dean W. DeChambeau
Senior Marketing Manager: Nathan Schultz
Manufacturing Buyer: Amy Bacus
Composition: Ann Marie Lemoine
Text Design: Anne Spencer
Cover Design: Kristin E. Ohlin
Printing and Binding: Courier Kendallville
Cover and Text Printing: Courier Kendallville
Cover Image: © Picture Quest

Library of Congress Cataloging-in-Publication Data
Chiras, Daniel D.
Human body systems : organization and structure / Daniel D. Chiras.
p. ; cm.
ISBN 0-7637-2356-8
1. Human biology. 2. Human physiology. 3. Human anatomy. I. Title.
[DNLM: 1. Anatomy. QS 4 C541h 2003]
QP34.5 .C4855 2003
612--dc21

2002043098

Printed in the United States of America
06 05 04 03 02 10 9 8 7 6 5 4 3 2 1

Contents

Introduction

Human Body Systems: Structure, Function and Environment is written for students who want to learn more about their bodies. This book focuses primarily on the organ systems that constitute the body. You'll learn the main parts of each system and how they operate. You'll also learn what happens when things go wrong.

In writing this book, I've attempted to distill information about human anatomy and physiology to the very basics, selecting information that should be most useful to you. There's so much useful information in here, in fact, that this book could be a valuable reference throughout your life. You may want to hang on to this book and refer to it from time to time — for example, when you or a family member gets sick and you want to understand what's wrong and what can be done to get well. If you want to learn more, you can always refer to the big sister of this book, *Human Biology: Health, Homeostasis, and the Environment, Fourth Edition.*

Human Body Systems: Structure, Function and Environment also discusses the biological systems, called ecosystems, that we live in. These systems provide us with a wealth of resources and truly are the life support system of the planet. Our own well-being depends mightily on their health. The very last chapter of the book tackles major environmental issues, problems threatening the health and well-being of the planet's life support systems. This, too, is valuable information that could prove useful throughout your life. As you read about issues in the newspaper or hear reports on television or the radio, you can refer to the last two chapters to gather more information.

An Introduction to Structure and Function

The human body is a marvel of structure and function made from billions and billions of cells. These cells combine to form tissues, such as muscle and connective tissue. Tissues, in turn, merge to form organs. Organs that perform related functions form organ systems of the body.

Body Tissues

The human body consists of four tissue types: (1) epithelial (ep-eh-theel-E-al), (2) connective, (3) muscle, and (4) nervous. These are called the **primary tissues**.

Each primary tissue consists of two or three subtypes. Muscle tissue, for instance, comes in three varieties—these will be discussed shortly. The primary tissues exist in all organs but in varying amounts. The wall of the stomach, for example, consists of a single layer of epithelial (ep-eh-THEEL-ee-ill) cells forming the inner lining (FIGURE 1-1). It's a protective coating. Just beneath the epithelial lining is a layer of connective tissue. It holds things together. Beneath that is a thick sheet of smooth muscle cells, which form the bulk of the stomach wall. Nerves enter with the blood vessels and control the flow of blood. Let's take a look at each primary tissue.

Epithelium. Most epithelium forms internal and external linings of various organs of the body. However, some epithelial tissue forms important glands. The epithelium forming linings or coverings typically consist of one or more layers of cells (FIGURE 1-2). Their appearance often reflects their function. The epithelial lining of the intestine, for example, consists of a single layer of cells, ideal for absorbing food molecules. In contrast, skin consists of many layers, providing much-needed protection.

The epithelium that forms many of the glands of the body does so during embryonic development. Some of these glands remain connected by hollow ducts to the epithelial layer from which they were formed. The glands release their products into the ducts which deliver them to the proper location. In humans, sweat glands in the skin produce a clear, watery fluid that is released onto the surface of the skin through ducts. This fluid evaporates from the skin and cools the body.

Some glands break off completely during embryonic development. Their products are released into blood vessels inside the gland. They are then transported throughout the body in the bloodstream. A good example is the thyroid gland in the neck. It produces hormones that stimulate metabolism.

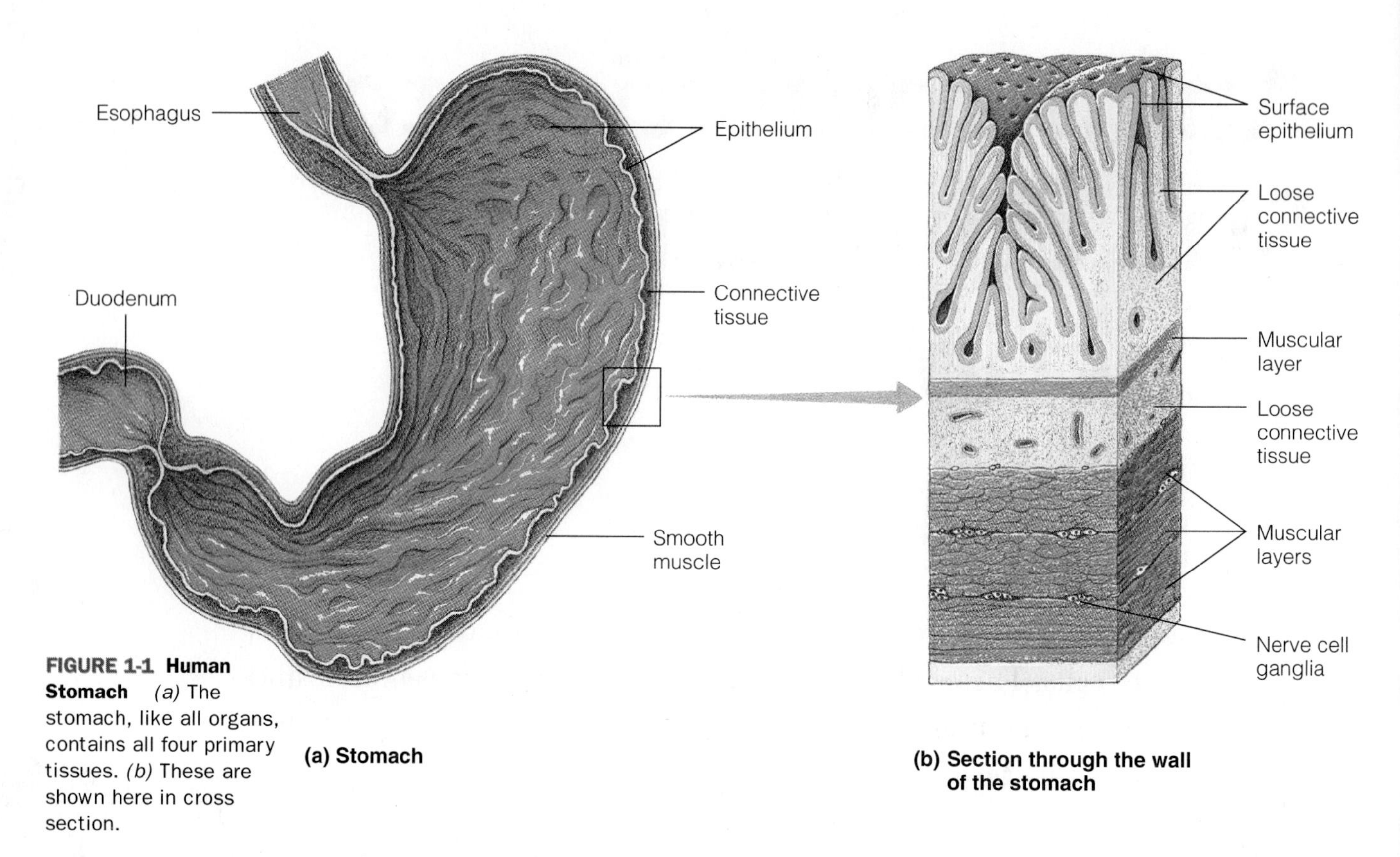

FIGURE 1-1 Human Stomach *(a)* The stomach, like all organs, contains all four primary tissues. *(b)* These are shown here in cross section.

Connective Tissue. As the name implies, connective tissue binds cells and other tissues together. Present in all organs in varying amounts, connective tissue consists of two basic components: cells and varying amounts of extracellular material, such as fibers.

The body contains several types of connective tissue, each with specific functions (FIGURE 1-3). Some connective tissue consists of a loosely knit mesh of fibers and cells. It serves as a packing material around cells. Fat cells are often located in this type of connective tissue.

Fat cells can contain huge fat globules that press the cytoplasm and the nucleus to the outside (Figure 1-3). Fat cells occur alone or in groups of varying size. When a loose connective tissue contains many fat cells it is called fat or **adipose tissue.** Fat tissue is an important storage depot for energy and also provides insulation against cold.

Also located in loose connective tissue is a group of cells that guard against bacterial infections. They are known as macrophages, a fancy name for "big eaters." These cells get this name because they can gobble up bacteria that enter tissues, and thus provide protection from infection.

Other connective tissue consists primarily of densely packed fibers. It is found in ligaments and tendons. **Ligaments** are structures that reinforce joints. They join bones to bones at joints. **Tendons** connect muscles to the bones of the body. When the muscle contracts, the tendon pulls on the bone, causing movement.

Specialized Connective Tissues. The human body also contains three types of specialized connective tissue: cartilage, bone, and blood.

Cartilage is found in many parts of the body—for example, the tip of the nose, the ears, the ends of the long bones, and between the bones of the spine. Cartilage has no blood vessels and must be nourished by nutrients that come from blood vessels surrounding this tissue. For this reason, cartilage heals very slowly, if at all, when damaged.

Bone is dynamic, living tissue that forms the internal framework of our bodies—our skeleton. Some bones provide protection to internal organs, such as the brain, heart, and lungs. In addition, bone acts as a calcium reserve that helps us maintain optimal blood calcium levels. Calcium is required for many body functions, such as muscle contraction, normal nerve functioning, and blood clotting.

Blood. Blood consists of two types of cells: red blood cells and white blood cells. Blood also contains cell fragments called *platelets*. The cells

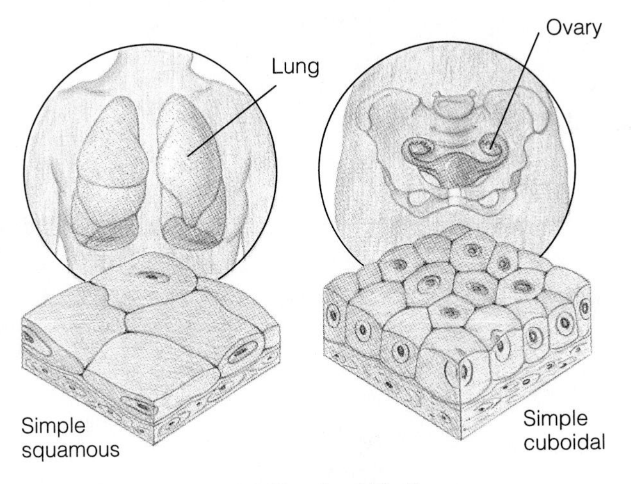

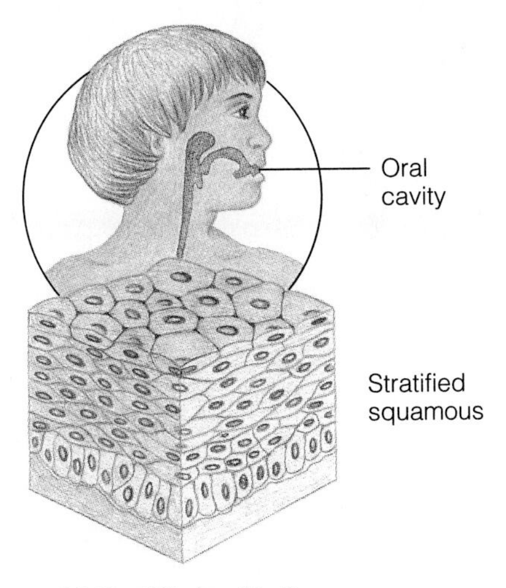

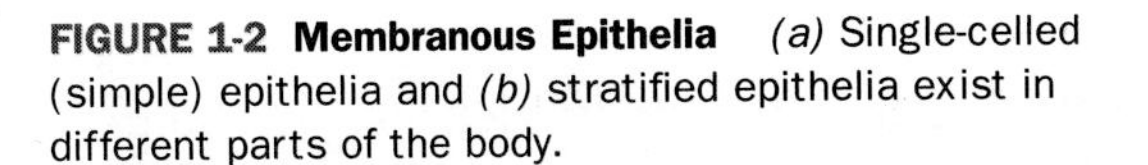
FIGURE 1-2 Membranous Epithelia *(a)* Single-celled (simple) epithelia and *(b)* stratified epithelia exist in different parts of the body.

and platelets are suspended in a fluid called plasma.

Red blood cells transport oxygen to the cells of the body from the lungs and transport waste carbon dioxide in the other direction. **White blood cells** are involved in fighting infections and cancer. **Platelets** play a key role in blood clotting.

Muscle. Muscle gets its name from the Latin word for "mouse" (mus). Early observers likened the contracting muscle of the biceps to a mouse moving under a carpet.

When muscle contracts, it causes body parts like the arms and fingers to move. Working in large numbers, muscle cells can create enormous forces. Muscles of the jaw, for instance, create a pressure of 200 pounds per square inch, forceful enough to snap off a finger. (Don't try this at home.) Muscle

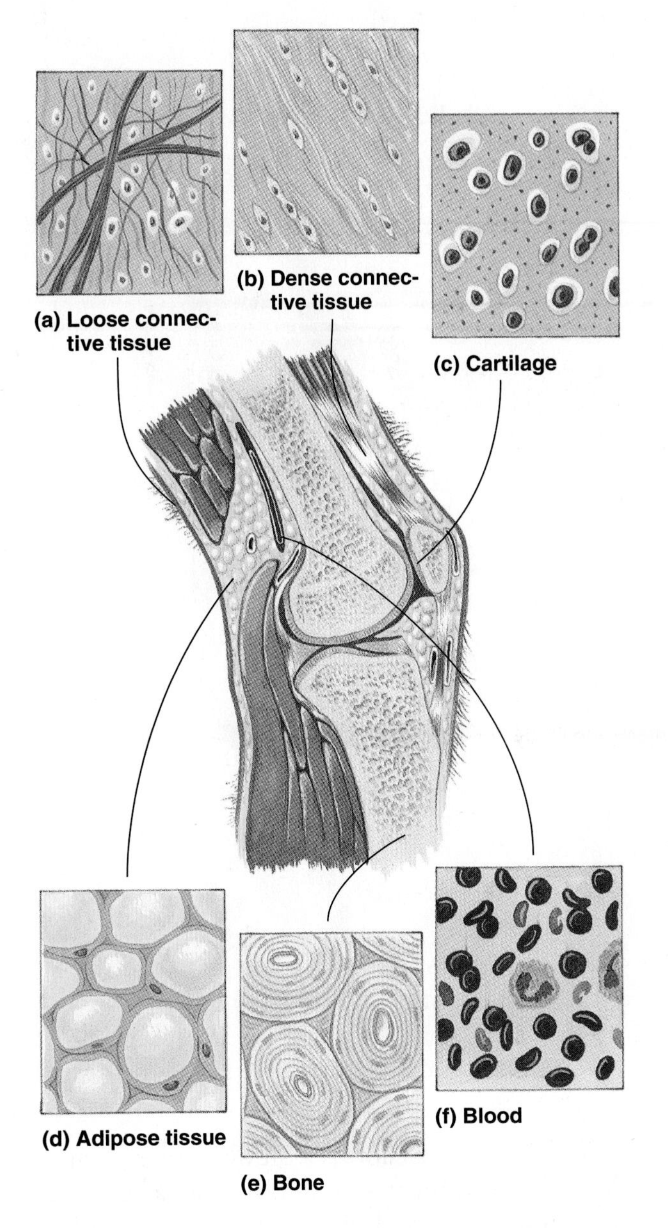

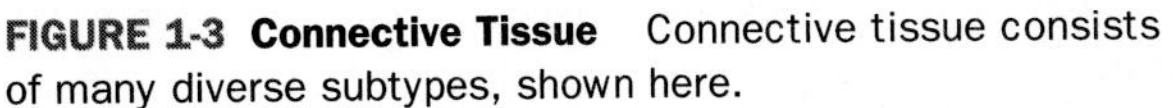
FIGURE 1-3 Connective Tissue Connective tissue consists of many diverse subtypes, shown here.

also propels food along the digestive tract and expels the fetus from the uterus during birth. Heart muscle contracts and pumps blood through the 50,000 miles of blood vessels in the body. Acting in smaller numbers, muscle cells are responsible for intricate movements, such as those required to play the piano or move the eyes.

As mentioned earlier, three types of muscle are found in humans: skeletal, cardiac, and smooth. The majority of the body's muscle is called **skeletal muscle**—so named because it is usually attached to the skeleton. Most skeletal muscle in the body is under voluntary control.

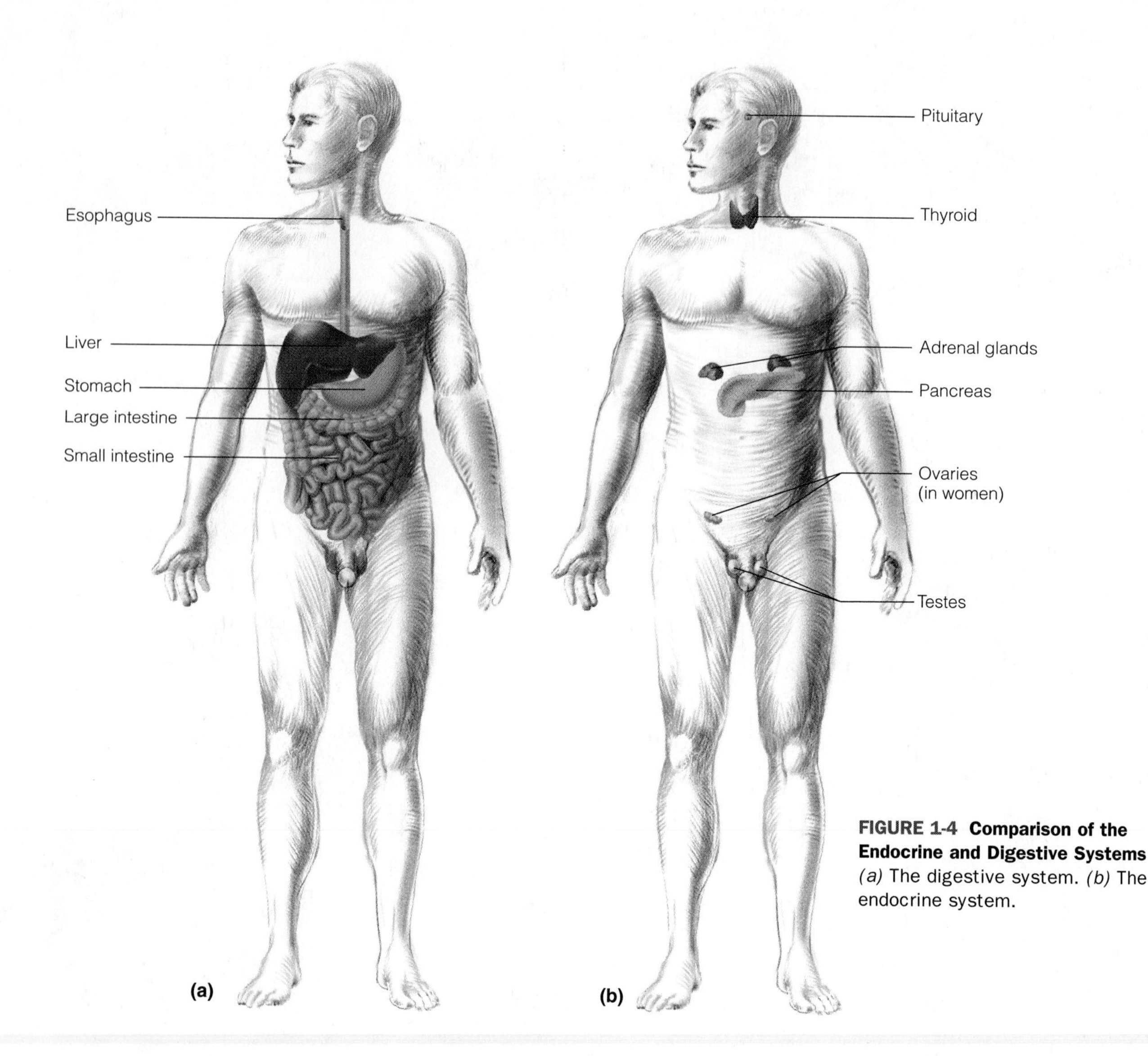

FIGURE 1-4 Comparison of the Endocrine and Digestive Systems *(a)* The digestive system. *(b)* The endocrine system.

Cardiac muscle looks a bit like skeletal muscle, but is involuntary—that is, it contracts without conscious control. It is found only in the walls of the heart.

Smooth muscle occurs alone or in small groups. Small rings of smooth muscle cells, for example, surround tiny blood vessels. When these cells contract, they shut off or reduce the supply of blood to tissues. Smooth muscle cells are most often arranged in sheets in the walls of organs, such as the stomach and intestines (Figure 1-1). Smooth muscle cells in the wall of the stomach churn the food, mixing the stomach contents, and force the liquefied food into the small intestine. Smooth muscle contractions also propel the food along the intestinal tract.

Nervous Tissue. Last but not least of the primary tissues is nervous tissue. Nervous tissue consists of nerve cells, or **neurons**, which are adapted to respond to stimuli, such as pain or temperature. Stimulation of these cells results in tiny impulses. Neurons transmit these impulses from one region of the body to another, helping control movements and many other functions.

The other type of cell found in nervous tissue is a kind of nervous system connective tissue. These cells are incapable of conducting impulses, but they perform other important functions. For example, they transport nutrients from blood vessels to neurons and form a barrier to many potentially harmful substances. Together, the neurons and their supporting cells form the brain, spinal cord, and nerves of the nervous system.

Organs and Organ Systems

Organs are structures in the body that have evolved to perform specific functions. Organs with similar function, like the heart and blood vessels, form **organ systems**.

As you will see in upcoming modules, components of an organ system are sometimes physically connected—as in the digestive system. In other cases, they are dispersed throughout the body—as in the endocrine system, the system that produces hormones (**FIGURE 1-4**). Some organs belong to more than one system.

Modules 2-15 describe the major organ systems and the functions they perform, paying special attention to their role in a process called *homeostasis* (hoe-mee-oh-stay-siss).

Principles of Homeostasis

The body's many systems perform specific functions that help us move about in our environment and perform work or play. Many systems also contribute to homeostasis. This process, as you will soon see, is essential for maintaining internal conditions vital to the survival of cells, tissues, and organs—and to our own survival as well.

Homeostasis is a state of relative internal constancy. Homeostasis is maintained by a process called negative feedback. **Negative feedback** is best illustrated by studying how a home heating system operates.

In the winter, the furnace of a house maintains a constant internal temperature, even though the outside temperature fluctuates dramatically. The furnace, in turn, is controlled by a thermostat, which monitors room temperature. When the indoor temperature falls below the desired setting, the thermostat sends an electric signal to the furnace, turning it on. Heat is generated by the furnace and is then distributed through the house, raising the room temperature. When the room temperature reaches the desired setting, the thermostat shuts the furnace off. This is a negative feedback mechanism.

Systems in the human body that maintain homeostasis operate similarly, but like the furnace they do not maintain absolute constancy. Rather, they maintain conditions such as body temperature within a given range around a set point.

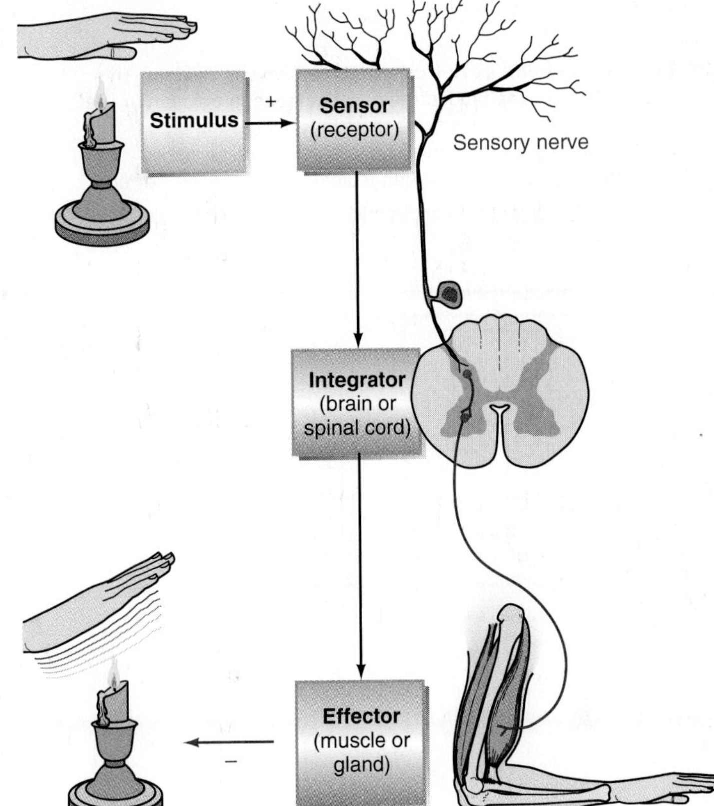

FIGURE 1-5 Nervous System Reflex Reflexes involve some kind of stimulus, a sensor, an integrator, and an effector. The sensor detects a change and sends a signal to the integrator, the brain or spinal cord, which then elicits a response in the effector organs. Negative feedback from the effector eliminates the stimulus.

As you will see in later modules, our bodies maintain fairly constant levels of a great many chemicals, such as hormones and nutrients. We also maintain relatively constant physical conditions, such as temperature and blood pressure. All of this is done automatically.

Homeostasis in the human body requires **sensors** that detect changes in internal conditions and **effectors** that correct these conditions. In your home, the thermostat is the sensor, and the furnace is the effector.

The human body contains many sensors. In your body, for instance, specially modified nerve cell endings in the skin detect temperature in the environment. The nerves send signals to the brain, alerting it to changes. The brain then sends signals to the body to correct matters—for example, in this instance, to increase heat production.

Most homeostatic mechanisms are **reflexes**—that is, automatic responses triggered by stimuli.

FIGURE 1-6 Endocrine Reflex Not Involving the Nervous System This reflex operates through the bloodstream. A decrease in calcium in the blood stimulates a series of reactions that restores normal blood calcium levels. A plus sign indicates that the stimulus increases parathyroid gland activity; a minus sign indicates the opposite effect.

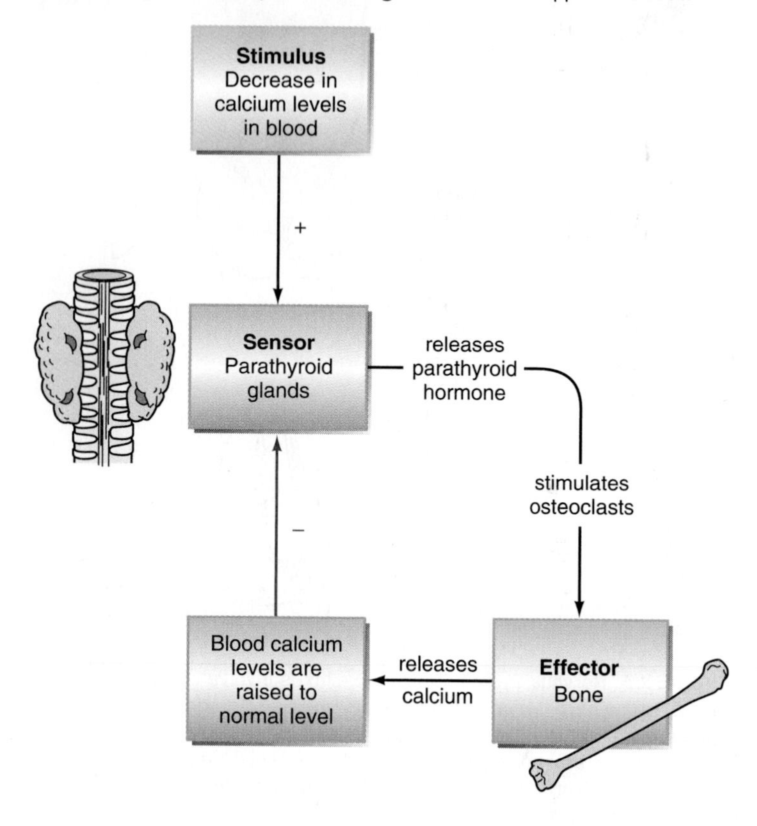

Two types of reflexes exist: nervous system and hormonal.

Homeostatic systems controlled by the nervous system are fairly straightforward. As shown in **FIGURE 1-5**, heat from a candle is detected by sensors in the skin. These receptors send nerve impulses to the spinal cord via sensory nerves. The spinal cord, in turn, sends a signal to pull the hand away from the flame.

Hormones are chemical substances produced by certain glands in the body, known as *endocrine glands*. Released into the blood, hormones travel to distant sites where they cause some kind of change. But hormones don't just pour out of endocrine glands in a constant stream. They are released only under certain circumstances, usually as part of a chemical reflex. For example, when calcium levels in the blood fall below a certain concentration, tiny glands in the neck, known as the *parathyroid glands* (pair-ah-THIGH-roid), release a hormone (parathyroid hormone). It travels in the blood to the bone (**FIGURE 1-6**). Here it stimulates certain cells that eat away at the bone. This releases calcium stored in the bone and raises the calcium level of the blood. Calcium levels return to normal.

Health and Homeostasis

Human health is dependent on homeostasis. When homeostasis is altered, we can become sick, even die. Homeostasis, in turn, requires a healthy, clean environment.

FIGURE 1-7 Health, Homeostasis, and the Environment Human health is dependent on maintaining homeostasis. Homeostasis, however, is affected by the condition of our environment. Stress, pollution, noise, and other environmental factors upset the function of cells and body systems, thus upsetting homeostasis and human health.

Diet and Lifestyle

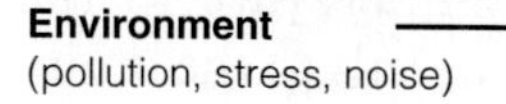

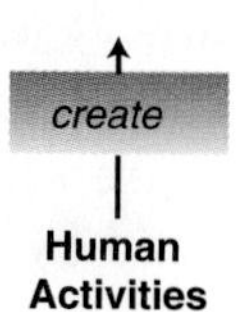

Diet, Lifestyle, and Environmental conditions affect the function of body cells.

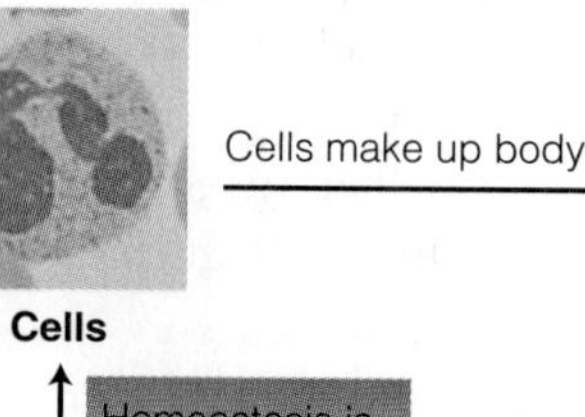

Cells

Cells make up body.

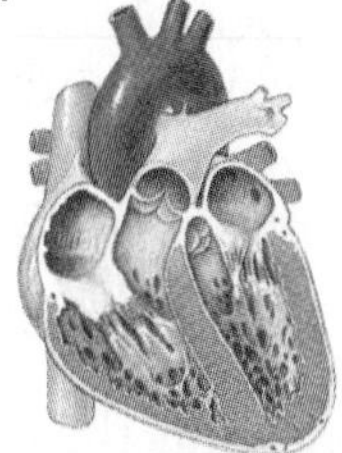

Organ Systems

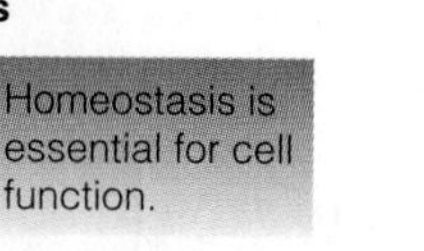

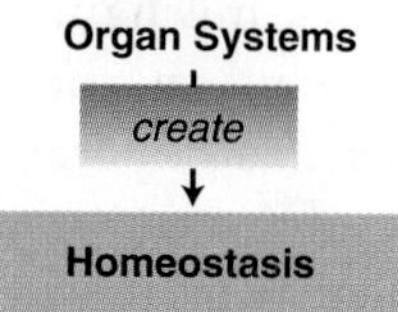

The relationship between homeostasis and health is shown in **FIGURE 1-7**. Take a moment to study this diagram. Notice that the body systems (right side of diagram) maintain homeostasis. It is essential for proper cell function. Because cells make up body organs, their health is essential to our well-being and survival. Notice too that environmental factors, such as pollution, affect the function of body systems and cells in ways that can upset the internal balance, thus altering human health.

Biological Rhythms

The previous discussion may have given you the impression that homeostasis establishes an unwavering condition of stability that remains more or less the same, day after day, year after year. In truth, many body processes undergo daily or monthly changes.

Body temperature varies during a 24-hour period by as much as a half degree Celsius. Blood pressure may change by as much as 20%, and the number of white blood cells, which fight infection, can vary by 50% during the day. Alertness also varies considerably. About 1:00 P.M. each day, for instance, most people go through a slump. For most of us, activity and alertness peak early in the evening, making this an excellent time to study. Daily cycles are natural body rhythms linked to the 24-hour day-night cycle.

Many hormones follow daily cycles. The male sex hormone testosterone, for example, is released in a 24-hour cycle. The highest levels occur in the night, particularly during dream sleep, also known as **REM sleep**. Dream sleep occurs primarily in the early morning hours.

Not all cycles occur over 24 hours, however. Some can be much longer. The menstrual cycle, for instance, is a recurring series of events in the reproductive functions of women. It lasts, on average, 28 days. During the menstrual cycle, levels of the female sex hormone estrogen undergo dramatic shifts. Estrogen concentrations in the blood are low at the beginning of each cycle and peak on day 14, when ovulation (release of the egg) normally occurs. Throughout the remaining 14 days, estrogen levels are rather high. They drop off again when a new cycle begins. Estrogen levels follow this cycle month after month in women of reproductive age.

The important point here is that the body is not static. Although many chemical substances are held within a fairly narrow range by homeostatic mechanisms, others fluctuate widely in normal cycles. Over the long run, these changes are quite predictable.

Internal biological rhythms are controlled by the brain. Just how the body controls its many internal rhythms remains a mystery. Research suggests a clump of nerve cells in the base of the brain in a region called the *hypothalamus*. It may regulate other control centers and is often referred to as the "master clock."

Ultimate control of the master clock is thought to occur in a gland in the brain known as the *pineal gland* (PIE-knee-al). It secretes a hormone thought to keep the master clock in sync with the 24-hour day-night cycle.

The study of biological rhythms is a fascinating field that has yielded some important insights. One practical application is a better understanding of jet lag—that drowsy, uncomfortable feeling people get from the disruption of sleeping patterns caused by long-distance jet travel. Studies suggest that jet lag occurs when the body's biological clock is thrown out of synchrony with the day-night cycle of a traveler's new surroundings. A business woman who travels from Los Angeles to New York, for instance, may be wide awake at 10:00 P.M. New York time because her body is still on Los Angeles time—3 hours earlier. When the alarm goes off at 6:00 A.M., our weary traveler crawls out of bed exhausted. As far as her body is concerned, it is 3:00 A.M. Los Angeles time.

Probably the greatest disrupter of our natural body rhythms is the variable work schedule, which is surprisingly common among industrialized nations. Workers on alternating shifts suffer from a higher incidence of ulcers, insomnia, irritability, depression, and tension than workers on regular shifts. Making matters worse, tired, irritable workers may suffer from impaired judgment, posing a threat not only to themselves but also to society. Fortunately, researchers are finding ways to reset the biological clock, which could reduce the problems shift workers face and thereby increase work performance, particularly of those on the graveyard shift.

Nutrition

Despite the increased emphasis on good nutrition today, research suggests that most Americans pay little attention to their diet. To perform and feel our very best, though, we must eat well. Eating a good diet provides the energy and nutrients cells, tissues, and organs of our bodies require, and helps us to maintain a healthy weight. It also helps us feel better and live longer lives with fewer diseases.

A **balanced diet** is attained by eating a variety of foods. In 1992, the U.S. Department of Agriculture released a helpful tool called a **food pyramid**, designed to assist those interested in eating healthy. The food pyramid, shown in **FIGURE 2-1**, places foods in six major groups. As you can see, it also indicates how many servings you need to eat from each group, each day, for proper nutrition. Take a moment to study the pyramid, starting at the bottom. As you can see, a healthy diet consists of plenty of breads, cereals, rice, and pasta—about 6 to 11 servings per day. The next largest group, the fruits and vegetables, 5 to 9 servings per day. Meat and milk products are next, but are required in lesser amounts. Fat is found in many foods, including meats, vegetables, and milk products. They've been singled out, nonethless, with the advice to "use sparingly."

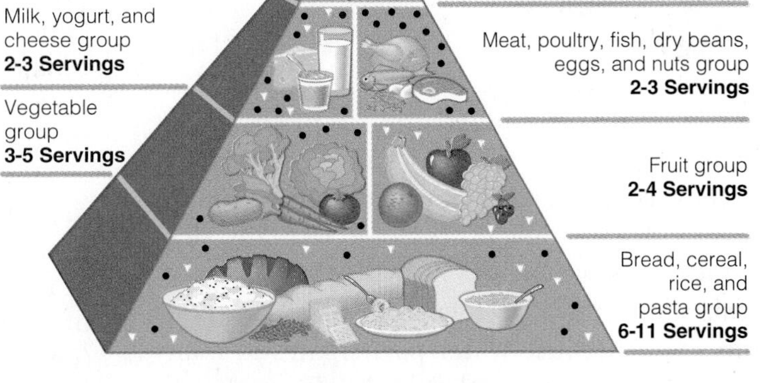

FIGURE 2-1 Food Guide Pyramid The food pyramid divides food into six groups. Recommended daily intake is shown for each group.

So how does your diet compare? If you are like most Americans, you probably go wrong by eating too few fruits and vegetables and too many meats, milk products, and fats. By adjusting your diet very slightly—for example, eat some fruit a couple times a day, have a salad every day and cut back on fatty foods—you can make important strides in achieving a more healthy diet with a greatly lowered risk of heart attack, stroke, obesity, and diabetes.

The nutrients we need to survive and prosper physically and mentally fit into two broad categories: macronutrients and micronutrients. **TABLE 2-1** lists the basic nutrients and includes some of the foods and beverages that provide them.

TABLE 2-1

Macronutrients and Micronutrients

Nutrients	Foods Containing Them
Macronutrients	
Water	All drinks and many foods
Amino acids and proteins	Milk and milk products such as cheese; soy products such as tofu and soy milk; meat, and eggs
Lipids	Milk and milk products, meats, eggs, nuts, oils, and seeds
Carbohydrates	Breads, pastas, cereals, sweets
Micronutrients	
Vitamins	Many vegetables, meats, and fruits
Minerals	Many vegetables, meats, fruits, nuts, and seeds

Macronutrients: Water, Lipids, Carbohydrates, and Protein

Macronutrients are required in relatively large amounts, hence their name. This group includes four substances: water, carbohydrates, lipids, and proteins

Water

Water is one of the most important of all the substances we ingest and is supplied in many different ways. For example, water is ingested in the liquids we drink, and it is present in virtually all of the solid foods we eat.

Maintaining the proper level of water in the body is important because water participates in many chemical reactions in the body. Even energy production can be affected by inadequate water intake. Adequate water intake helps to stabilize body temperature and assists in keeping the level of toxic waste products in the blood and fluid surrounding the cells of our bodies from getting too high. Proper water intake keeps the urine from becoming too concentrated. If you don't consume enough water, calcium and other chemicals in the urine may rise, increasing your chance of developing a kidney stone. Kidney stones are deposits made of calcium and other materials that can block the flow of urine, causing extensive damage to the kidneys and pain.

Carbohydrates

Carbohydrates are a group of organic compounds that includes such well-known examples as table sugar (sucrose), blood sugar (glucose), and starch. Carbohydrates serve many functions, one of the most important is as a source of energy.

Glucose (glew-kose) is one of the most important carbohydrates. In the body, it is broken down to make energy.

People derive the glucose they need primarily from starch, a carbohydrate produced by plants. Starch from plants is used to produce various foods such as bread and pasta. Many vegetables such as corn and potatoes also contain starch.

In the digestive system, starch is broken down into glucose molecules. Glucose then enters the blood stream. Here it circulates throughout the body, providing a source of energy for the cells.

Excess glucose taken in during a meal is stored in the liver and skeletal muscles. Here, glucose molecules are combined to form a long-chained polysaccharide called **glycogen** (gly-ko-gin). What is glycogen used for?

Glycogen supplies glucose in periods between meals. Liver stores of glycogen, for instance, are broken down into glucose molecules after meals. Glucose then enters the blood stream and feeds the cells of our body. Glycogen in muscles is broken down when we exercise to produce glucose required to supply energy, but only to muscle cells.

Humans need a continuous supply of glucose to produce energy, even the most ardent couch potatoes! Energy is used to carry out the thousands of functions at the cellular level. In fact, 70%-80% of the total energy required by sedentary individuals (people who get little exercise) is used to perform very basic functions—metabolism, food digestion, absorption, and so on. The remaining energy is used for body movements such as walking and talking. The more active you are, however, the more energy is needed for muscle contraction.

Cellulose is another important carbohydrate. Made from glucose molecules, cellulose is one form of dietary fiber known as *water-insoluble fiber*. This means that it cannot dissolve in water.

Cellulose is present in many vegetables such as celery and carrots as well as many grain products such as cereal and breads. Interestingly, humans do not contain the enzymes in their digestive tracks required to digest cellulose. So why is it so important?

Although cellulose cannot be digested and used as an energy source, these undigested fibers help keep feces more liquid by retaining water in the large intestine. This, in turn, permits the passage of feces through the large intestine, preventing constipation. It may also reduce one's chances of getting colon cancer.

A good diet also contains a fair amount of water-soluble fiber. Water-soluble fiber is provided by fruits, vegetables, and some grains—including apples, bananas, carrots, and oats. Numerous research studies show that water-soluble fiber in fruits and vegetables helps lower blood cholesterol. Researchers think that they bind to cholesterol in our intestines, preventing it from being absorbed into the blood.

Lipids

Lipids are fats and oils found in plants and animals. They serve a variety of purposes. Some are energy-rich molecules that can be stored in the body for times of need. Still other lipids are structural components of cells.

One of the most common lipids is the **triglycerides**. They form animal fats and vegetable oils. When broken down by the body, these molecules release energy that can be used in cells. In fact, they yield more than twice as much energy as carbohydrates such as glucose.

In humans and many other mammals, triglycerides are stored in fat cells under the skin and in other locations (FIGURE 2-2). In adult humans, triglycerides provide about half of the energy we require when sitting around or hanging out on the couch, with glucose providing most of the rest. During moderate (aerobic) exercise, triglycerides provide a larger proportion of the body's energy demand, explaining one reason why aerobic exercise such as jogging and bicycling helps people lose weight.

As shown in FIGURE 2-3, each triglyceride molecule contains three fatty acids. Fatty acids in triglycerides from animals are known as *saturated fatty acids*. Consuming large quantities of fatty acids in animal fat increases one's risk of developing atherosclerosis (AH-ther-oh-skler-OH-siss).

Atherosclerosis is a disease that results from a buildup of cholesterol deposits, called plaque, on the walls of arteries (FIGURE 2-4). Plaque limits blood flow to vital organs such as the heart and brain, and can lead to heart attacks and strokes. Steak, hamburgers, cheese, chicken with its skin, bacon, whole milk, and many common snack foods contain lots of saturated fat.

Atherosclerosis is more common in sedentary people and smokers. Some individuals are more

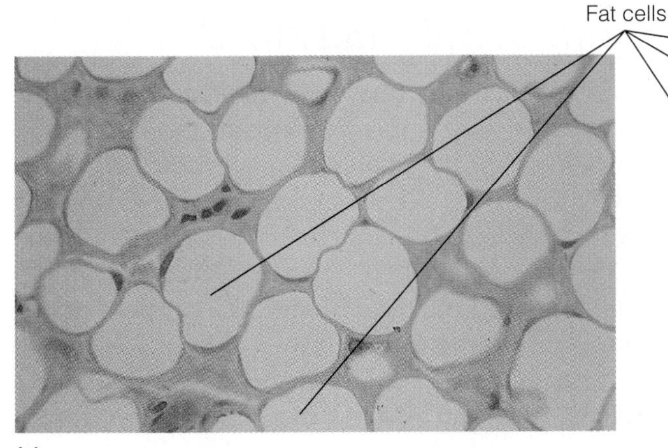

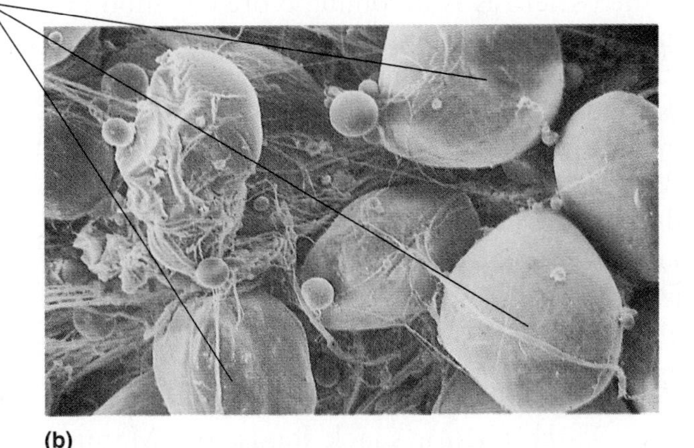

FIGURE 2-2 Fat Cells *(a)* A light micrograph of fat tissue. The clear areas are regions where the fat has dissolved during tissue preparation. Notice that the cytoplasm is reduced to a narrow region just beneath the plasma membrane. *(b)* A three-dimensional view of fat cells taken by an electron microscope.

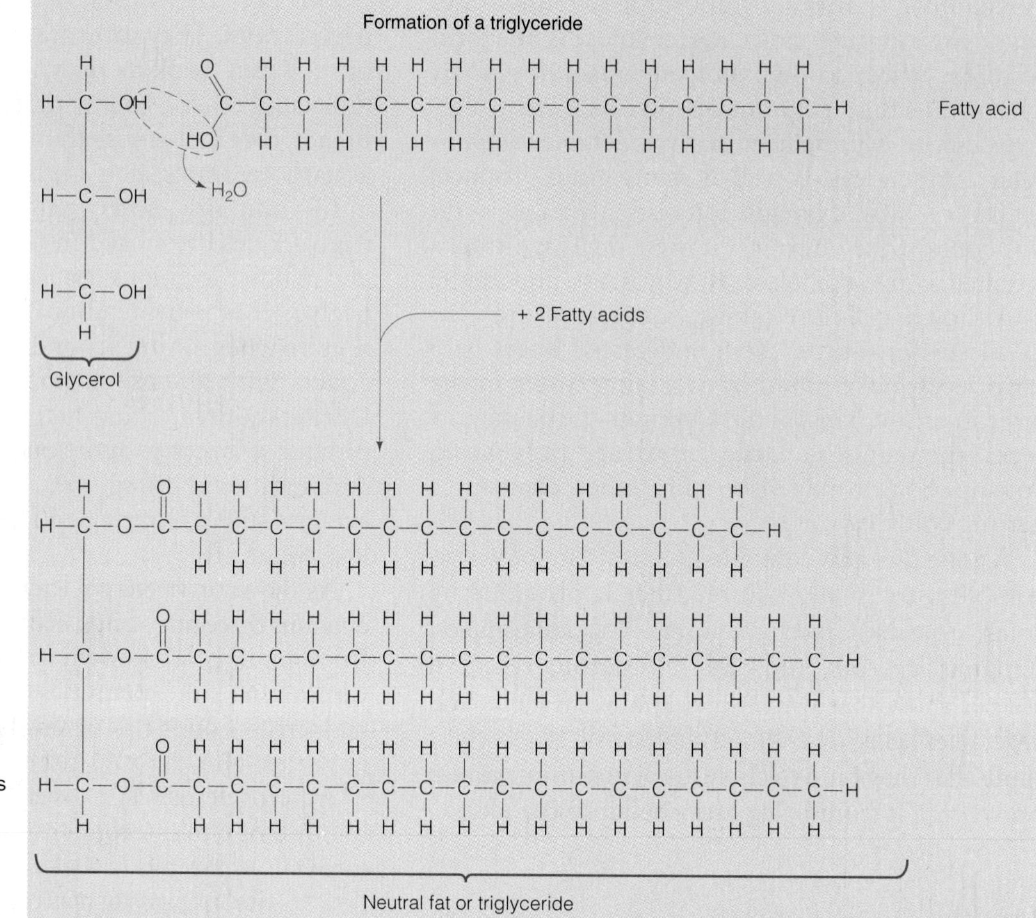

FIGURE 2-3 Triglycerides The triglycerides are the fats and oils. Triglycerides consist of glycerol and three fatty acids, covalently bonded as shown.

likely to have high levels of cholesterol in their blood and therefore more likely to develop atherosclerosis because of their genetics. That is, they contain a gene they inherited from their parents that causes the liver to produce abnormally high levels of cholesterol.

Atherosclerosis is becoming more common in adults in the United States and other industrial nations due to our high-fat diets and sedentary lifestyle. Signs of atherosclerosis are appearing even in people in their teens and twenties.

Lowering the level of saturated fat in the diet reduces cholesterol levels in the blood and reduces atherosclerosis. To do so requires a switch from whole milk to low-fat milk, reductions in the consumption of red meat, trimming fat from chicken and other meats, cutting down on cheese, and cooking with vegetable oil instead of animal fat (lard). Watch out for snack foods too. Manufacturers list the total fat and saturated fat content of all packaged foods.

Steroids. Steroids are a group of lipids that serve many purposes. One of the best-known steroids is cholesterol (**FIGURE 2-5**). **Cholesterol** is a component of the cell membranes and is used to make other steroids, such as vitamin D and the sex hormones in men and women.

Amino Acids and Protein

Amino acids and **proteins** are important nutrients. In healthy, well-nourished individuals, protein in our diets is broken down to produce amino acids. They are then used to make protein within our bodies, for example, enzymes and hormones. Many structural components such as collagen, which makes up our hair and fingernails and part of our bones, are proteins.

Proteins in the human body contain 20 different amino acids—all of which can be provided from the diet. The body, however, is capable of making 12 of the amino acids from nitrogen and smaller molecules derived from carbohydrates and fats. Thus, if they are not present in the diet, they can be made. Those amino acids that are not made in

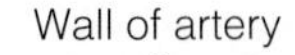
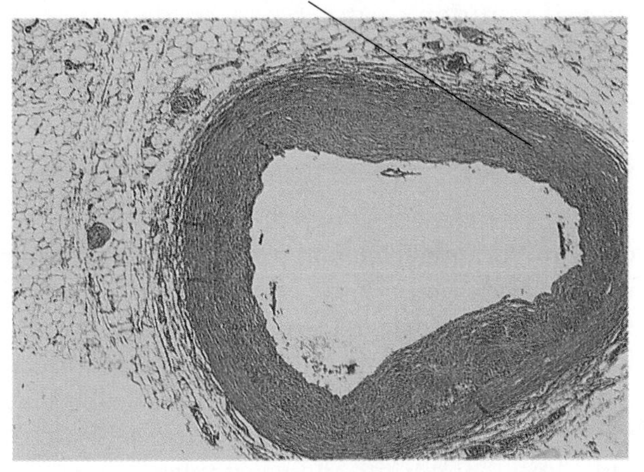

(a) Normal artery

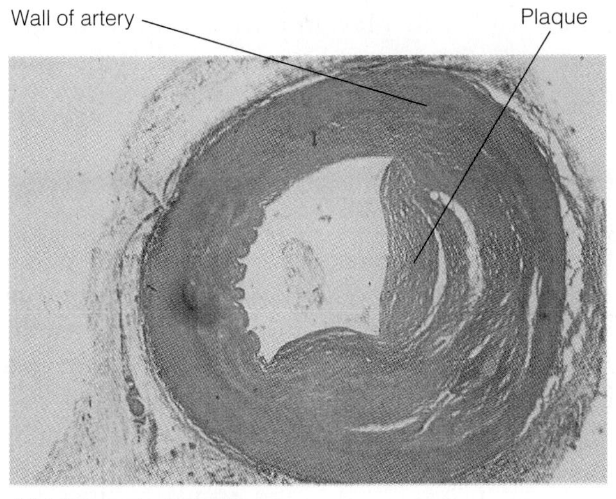

(b) Atherosclerotic artery

FIGURE 2-4 Atherosclerosis These cross sections of *(a)* a normal artery and *(b)* a diseased artery show how atherosclerotic plaque can obstruct blood flow.

sufficient amounts, however, must be provided in the diet. These amino acids are called **essential amino acids**. A deficiency of even one of the essential amino acids can cause severe health problems.

Some foods such as milk, eggs, meat, fish, poultry, cheese, and soy products, including soy milk and tofu, contain all of the amino acids we need. Other foods such as nuts, seeds, grains, most legumes (peas and beans), and vegetables are rich in protein but lack some of the amino acids. When eating a diet rich in grains, nuts and cereals, you need to combine foods from each group to obtain all of the essential amino acids you need (**FIGURE 2-6**).

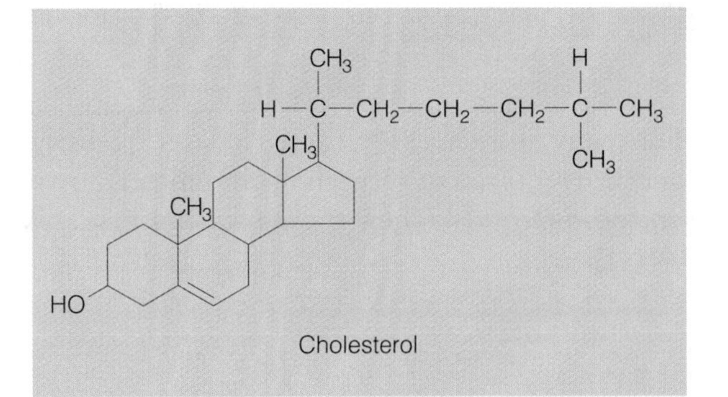

FIGURE 2-5 Steroids Steroids, such as cholesterol, consist of four rings joined together.

Overnutrition: Eating Too Much. While many of the world's people suffer from a lack of food, many residents of wealthier nations, consume too much. Excess food intake is called **overnutrition**. It can cause us to gain weight, which is then responsible for numerous problems, such as heart disease.

In the United States, 55% of all American adults are overweight. The percent of the population classified as severely overweight or obese has climbed from 15 to 23% since 1970. Even children are affected, with one of every five now classified as overweight or obese. Controlling weight by reducing food intake, eating healthier food, and exercise—even increasing walking—can help people lose weight and live healthier lives.

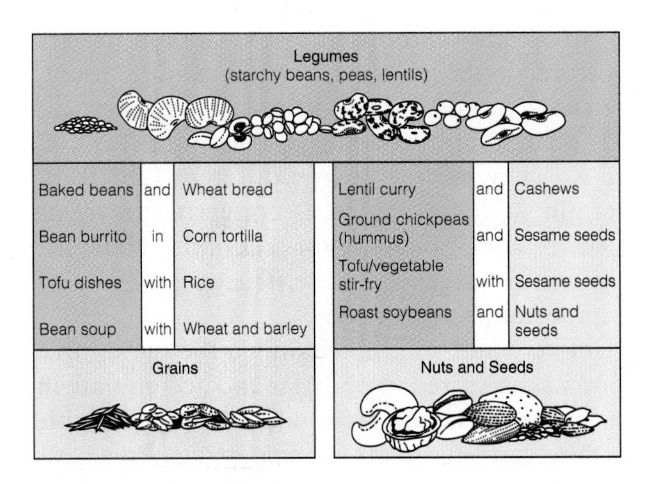

Legumes (starchy beans, peas, lentils)

Legumes		Grains	Legumes		Nuts and Seeds
Baked beans	and	Wheat bread	Lentil curry	and	Cashews
Bean burrito	in	Corn tortilla	Ground chickpeas (hummus)	and	Sesame seeds
Tofu dishes	with	Rice	Tofu/vegetable stir-fry	with	Sesame seeds
Bean soup	with	Wheat and barley	Roast soybeans	and	Nuts and seeds

FIGURE 2-6 Complementary Protein Sources By combining protein sources, a vegetarian who consumes no animal by-products can be assured of getting all of the amino acids needed. Legumes can be combined with foods made from grains or nuts and other seeds.

Micronutrients: Vitamins and Minerals

While many nutrients are required in large quantity, others are required in very small amounts. Known as *micronutrients*, they include vitamins and minerals.

Vitamins

Vitamins are a diverse group of organic compounds present in very small amounts in many of the foods we eat. The 13 known vitamins play an important role in many chemical reactions in the cells of the body. Because vitamins are recycled over and over, they are needed only in very small amounts.

Most vitamins are not made in the cells of the body or, if they can be made, are not produced in sufficient amounts, so they must be ingested in the food we eat.

Because vitamins are needed in almost all cells of the body, a dietary deficiency in just one vitamin can cause wide-ranging effects. Too much of a particular vitamin can also cause health problems.

Vitamins fall into two broad categories: water-soluble and fat-soluble. **Water-soluble vitamins** include vitamin C and eight different forms of vitamin B. Water-soluble vitamins generally work in conjunction with enzymes, promoting the cellular reactions that supply energy or synthesize cellular materials. Because they are water-soluble, they are readily eliminated by the kidneys and are not stored in the body in any appreciable amount. Even so, excesses of these vitamins can be toxic.

The **fat-soluble vitamins** are vitamins A, D, E, and K. They perform many different functions. Vitamin A, for example, is converted to light-sensitive pigments in receptor cells of the retina, the light-sensitive layer of the eye. These pigments play an important role in vision.

Unlike water-soluble vitamins, the fat-soluble vitamins are stored in body fat and accumulate in the fat reserves. The accumulation of fat-soluble vitamins can have many adverse effects. An excess of vitamin D, for example, can cause weight loss, nausea, and irritability.

Vitamin deficiencies, like dietary excesses, can lead to serious problems. A deficiency of vitamin D, for example, can produce **rickets** (RICK-its), a disease that results in bone deformities. Most dietitians recommend taking vitamin supplements only if you don't eat a balanced diet to avoid deficiencies and protect against overdoses.

Minerals

Humans require about two dozen minerals, such as calcium, sodium, iron, and potassium, to carry out normal body functions.

Minerals are derived from the food we eat and the beverages we drink. Calcium, for instance, is present in milk and milk products as well as dark green vegetables. Potassium is present in meats, milk, and many fruits and vegetables.

Minerals make up parts of our body such as the bones and also play important roles in body functions such as muscle contraction. Deficiencies can lead to serious problems, as can excesses.

Functional Foods

In recent years, scientists have discovered numerous chemicals in the foods we eat that help prevent disease. Foods containing them are called **functional foods** because they provide health benefits beyond basic nutrition.

The number of chemical substances that promote good health is really quite large. One widely publicized group of helpful chemicals is the **antioxidants**. They promote health by eliminating naturally occurring substances, the oxidants, in the blood that are involved in cholesterol build up in the walls of arteries. Oxidants may also contribute to cancer. Eating a diet rich in foods containing antioxidants may reduce heart attacks, stroke, and cancer.

Antioxidants are found in fruits and vegetables, soy milk, tofu, tea, and red wine. Even a few commercially available chocolate products may contain significant amounts of antioxidants. (At this writing, Mars, which makes Mars bars and other treats, is the only company that employs a process that preserves the beneficial antioxidants.)

In closing, good food leads to good health. This doesn't mean that you can't enjoy a banana split or a hamburger once in a while. You can. Just add exercise to your routine, like walking more each day, and add some fruits and vegetables. Eat less fat, and you're well on your way to good health.

The Digestive System

Food makes us what we are, but it is the digestive system that converts our food into usable molecules. Without these molecules, most foods would be of little value. The chemicals they contain simply can't be used by the body cells as they are.

Shown in FIGURE 3-1, the digestive system consists of numerous organs. Let's take a look at each part of the digestive system, starting in the mouth, to see how each one contributes to food digestion and absorption.

The Mouth

The **mouth** is a complex structure in which food is physically and chemically broken down. Food that enters the mouth is sliced and pulverized by the teeth in our jaws. This process results in the production of smaller, more digestible pieces.

As the food is chewed, it is mixed with saliva. **Saliva** is a watery secretion released by the salivary glands, located around the oral cavity. The release of saliva is triggered by the smell, feel, taste, and sometimes even by the thought of food.

Saliva helps to liquefy food in our mouths, making it easier to swallow. It dissolves food molecules so they can be tasted. It also begins to break down starch molecules with the aid of the enzyme **amylase** (AM-ah-lase), also produced by the salivary glands.

But saliva does more. It also cleanses the teeth, washing away bacteria and food particles. Because the release of saliva is greatly reduced during sleep, bacteria tend to accumulate on the surface of the teeth. Here they break down tiny food particles remaining in our mouths after meals. The breakdown of food particles, in turn, produces some foul-smelling chemicals that produce bad breath known as "dragon breath," or "morning mouth."

Controlling the bacteria that live on the teeth by regular brushing is important to prevent bad breath, but it also reduces cavities caused by bacteria. Bacteria secrete a sticky material called **plaque**. It adheres to the surface of our teeth, trapping the bacteria. The trapped bacteria then release small amounts of a weak acid that dissolves the hard outer coating of our teeth, the **enamel**. The acid forms small pits in the enamel, commonly referred to as **cavities**. If the cavity is left untreated, an entire tooth can be lost to decay.

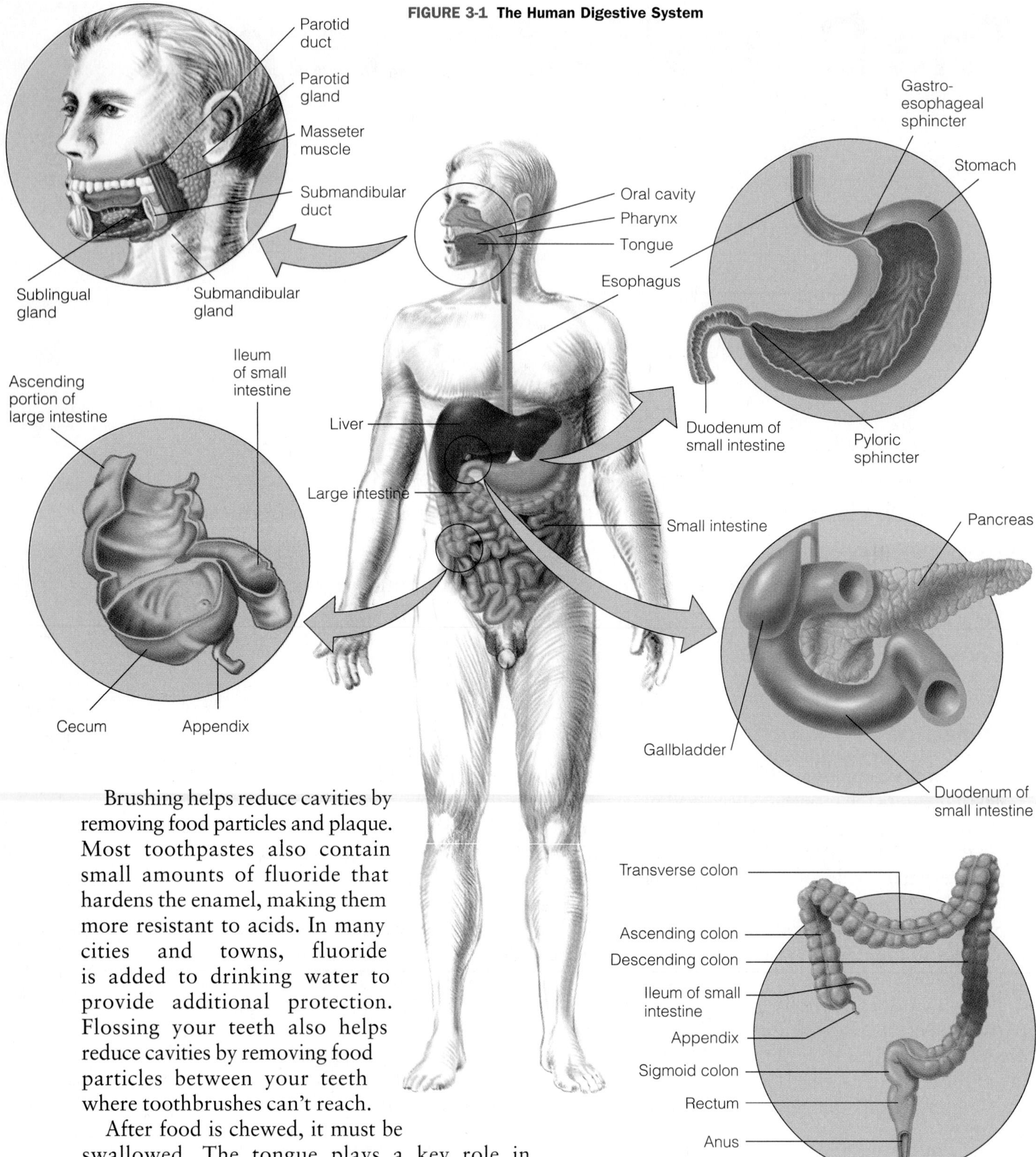

FIGURE 3-1 The Human Digestive System

Brushing helps reduce cavities by removing food particles and plaque. Most toothpastes also contain small amounts of fluoride that hardens the enamel, making them more resistant to acids. In many cities and towns, fluoride is added to drinking water to provide additional protection. Flossing your teeth also helps reduce cavities by removing food particles between your teeth where toothbrushes can't reach.

After food is chewed, it must be swallowed. The tongue plays a key role in swallowing by pushing food to the back of the oral cavity where it passes into the esophagus (ee-SOFF-ah-gus). The **esophagus** is a long muscular tube that leads to the stomach (Figure 3-1).

The **tongue** also aids in speech and contains taste receptors, or taste buds, on its upper surface (**FIGURE 3-2**). **Taste buds** are oval structures located in tiny bumps on the upper surface of the tongue. They are

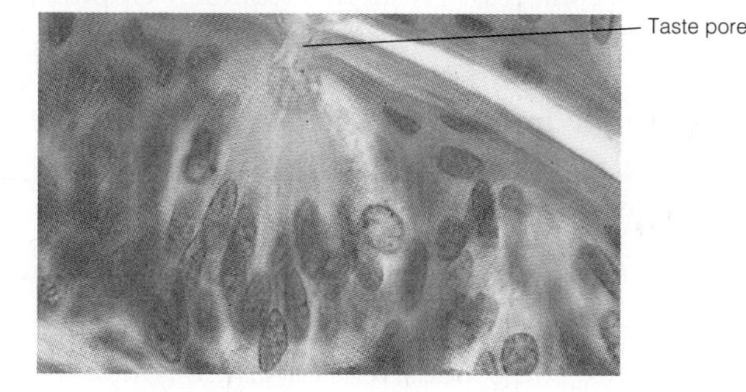

FIGURE 3-2 Taste Buds A photomicrograph of a taste bud.

stimulated by four basic flavors: sweet, sour, salty, and bitter. Various combinations of these flavors (combined with the odors we smell) give us a rich assortment of tastes.

Food propelled into the esophagus is prevented from entering the **trachea** (TRAY-key-ah), or windpipe, which carries air to the lungs and lies in front of the esophagus, by a flap of tissue called the **epiglottis** (ep-ah-GLOT-tis) (FIGURE 3-3). It acts like a trapdoor, closing off the trachea during swallowing (Module 7).

The Esophagus and Stomach

Swallowing is a voluntary action caused when the tongue pushes food into the back of the oral cavity. From this point on, however, movement of the food to the stomach is involuntary. The muscles in the wall of the esophagus push the food along (FIGURE 3-4).

In humans, the **stomach** (FIGURES 3-1 AND 3-5) lies on the left side of the abdominal cavity. When food reaches the stomach from the esophagus, a ring of muscles located at the attachment of the esophagus relaxes, allowing food to enter. The muscle then contacts after the food enters the stomach, preventing stomach acid from percolating upward. If the muscle fails to close, acid rising in the esophagus causes irritation, a condition known as *heartburn*.

Inside the stomach, food is liquefied by acidic secretions of tiny glands in the wall of the stomach. The food is churned by muscle contraction in the walls of the organ and is mixed with acid from the glands.

Combined with the liquid from salivary glands and the glands of the stomach, the food becomes a rather thin, watery paste. The stomach can hold 2-4 liters (2 quarts to 1 gallon) of liquefied food.

Contrary to what many think, very little digestion occurs in the stomach. The stomach's role is largely to prepare food for digestion by enzymes encountered in the next part of the digestive system, the small intestine. There are some exceptions to this rule, however. Protein is one of them.

Proteins are coagulated by stomach acid. Some protein-digesting enzyme is also present in the stomach. Produced by the glands that produce acid, this enzyme breaks the proteins down but only slightly. The protein fragments must be broken down further in the small intestine.

The stomach lining is protected from destruction by an alkaline secretion known as **mucus** (MEW-kuss). Produced by cells in the lining of the stomach, mucus protects the lining from acid and protein-digesting enzymes.

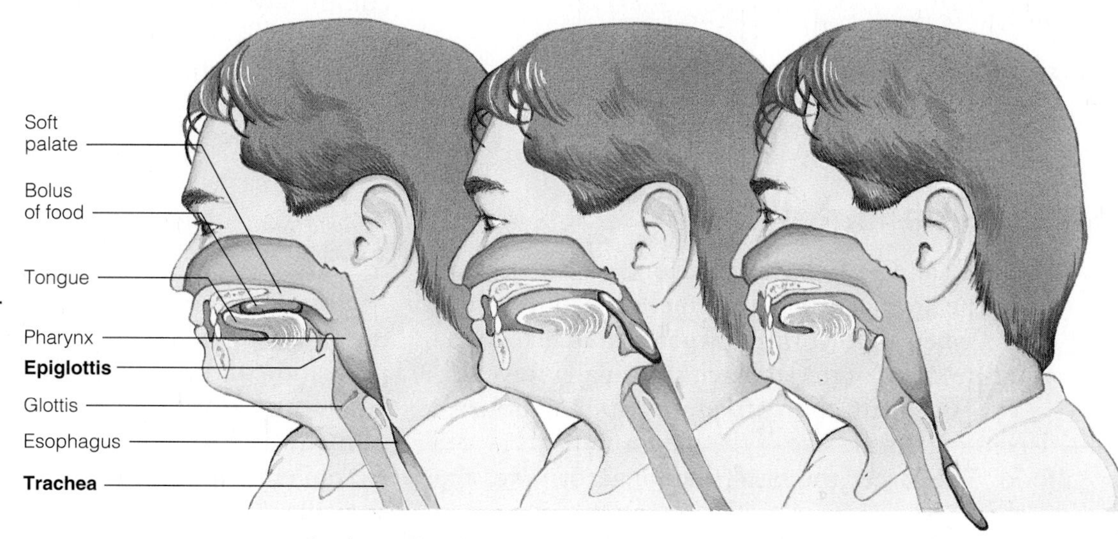

FIGURE 3-3 The Epiglottis This trapdoor prevents food from entering the trachea during swallowing. As illustrated, the trachea is lifted during swallowing, pushing against the epiglottis, which bends downward.

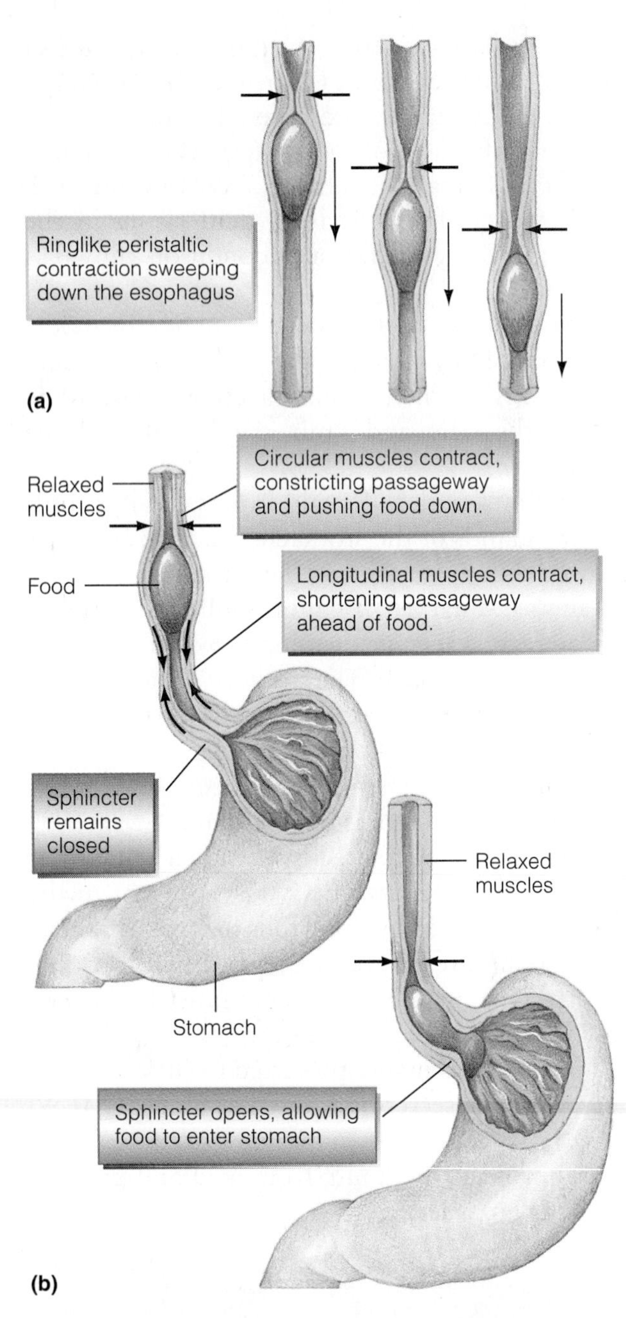

FIGURE 3-4 Peristalsis *(a)* Peristaltic contractions in the esophagus propel food into the stomach. *(b)* When food reaches the stomach, the gastroesophageal sphincter opens, allowing food to enter.

The Small Intestine

Liquefied and partially digested food is next ejected from the stomach into the small intestine. This occurs when another ring of smooth muscle relaxes at the junction of the stomach and small intestine.

The stomach contents empty in 2-6 hours, depending on the size of the meal and the type of food. The larger the meal, the longer it takes to empty. Solid foods (meat) empty slower than liquid foods (milk shakes). After the stomach empties, contractions in the walls of the organ continue. These contractions are felt as hunger pangs.

The **small intestine** is a coiled tube in the abdominal cavity about 6 meters (20 feet) long in adults (Figure 3-1). Inside the small intestine, large food molecules are broken into smaller ones. This process requires enzymes. Smaller food molecules can now pass through the lining of the small intestine into the bloodstream and the lymphatic system. The **lymphatic system**, discussed in Module 6, is a network of vessels that runs throughout the body. Many of these vessels carry excess tissue fluid from the tissues of the body to the circulatory system. Those found in the wall of the small intestine absorb fats and transport them to the bloodstream.

The digestion of food molecules inside the small intestine requires enzymes produced from two distinctly different sources: the pancreas, an organ that lies beneath the stomach, and the lining of the small intestine itself.

The **pancreas** is nestled in a loop formed by the first portion of the small intestine (**FIGURE 3-6**). The digestive enzymes of the pancreas flow through a large duct into the small intestine. The pancreas also produces sodium bicarbonate, a chemical that neutralizes the acids in the food produced by the glands of the stomach. This chemical protects the small intestine from stomach acid and creates an optimal environment for the function of pancreatic enzymes.

Pancreatic enzymes break down large molecules in food (proteins, starches, and so on), but not completely. The rest of the breakdown requires enzymes produced by the cells lining the small intestine. The final phase of digestion, therefore, occurs just before the nutrient is absorbed into the cells lining the small intestine.

The Liver

Digestion of food also requires the liver. Situated on the right side of the abdomen under the protection of the ribs, the **liver** is one of the largest and most versatile organs in the body. It performs as many as 500 different functions. The liver, for example, is one of the body's storage depots for

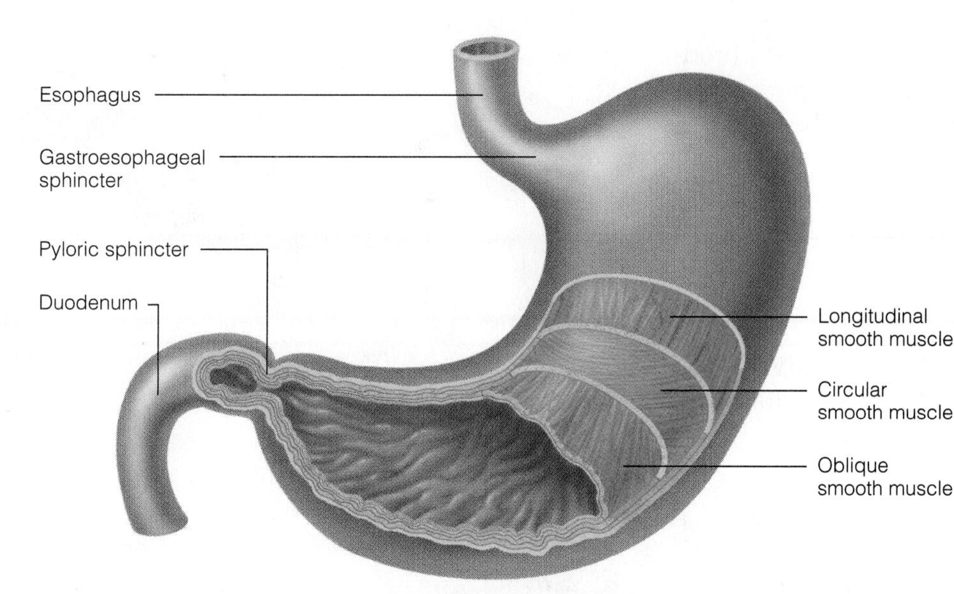

FIGURE 3-5 The Stomach The stomach lies in the abdominal cavity. In its wall are three layers of smooth muscle that mix the food and force it into the small intestine, where most digestion occurs. The gastroesophageal and pyloric sphincters control the inflow and outflow of food, respectively.

glucose. It also stores fats. By storing glucose and lipids and releasing them as they are needed, the liver helps ensure a constant supply of energy-rich molecules required by body cells. The liver also synthesizes some key proteins involved in blood clotting, and is an efficient detoxifier of potentially harmful chemicals such as nicotine, barbiturates, and alcohol.

The liver also plays a key role in the digestion of fats through the production of a fluid called bile. **Bile** is released into the small intestine when food is present. Bile contains **bile salts**. These steroids break fat globules into smaller ones. Smaller globules are then broken down further by fat-digesting enzymes.

Bile is produced by the cells of the liver and then stored in a small sac, the **gallbladder**, which is attached to the underside of the liver (**FIGURE 3-6**). The gallbladder concentrates the bile by removing water from it. When food is present in the small intestine, the gallbladder contracts, and bile flows out through a duct into the small intestine (Figure 3-6).

Bile flow to the small intestine may be blocked by **gallstones**, deposits of cholesterol and other materials in the gallbladder of some individuals. Gallstones may lodge in the ducts draining the organ, reducing—even blocking—the flow of bile. The lack of bile salts in the small intestine greatly reduces lipid digestion and absorption.

Gallstones occur more frequently in older, overweight individuals. When they cause problems, gallstones are usually surgically removed. This procedure requires that the entire gallbladder be removed.

Food Absorption by the Small Intestine

Once food molecules are digested, they are absorbed. Virtually all food digestion occurs in the small intestine. Absorption takes place in the small intestine as small food molecules, such as amino acids and glucose pass from the interior of the small

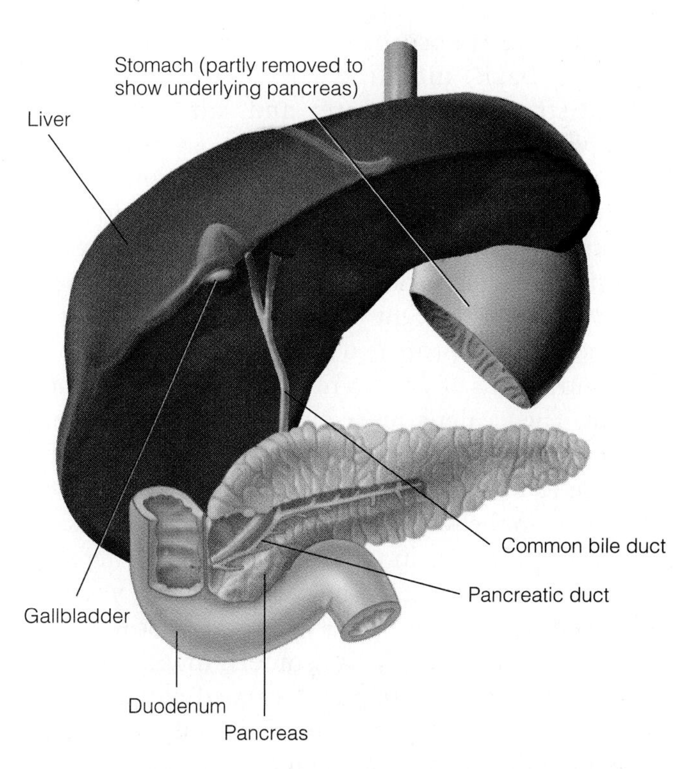

FIGURE 3-6 The Organs of Digestion The liver, gallbladder, and pancreas all play key roles in digestion. All empty by the common bile duct into the small intestine, in which digestion takes place.

intestine into the bloodstream or lymphatic system, located beneath the epithelium.

Numerous mechanisms are involved in absorption. Most nutrients pass through the lining of the intestine into the blood capillaries, tiny blood vessels, by diffusion.

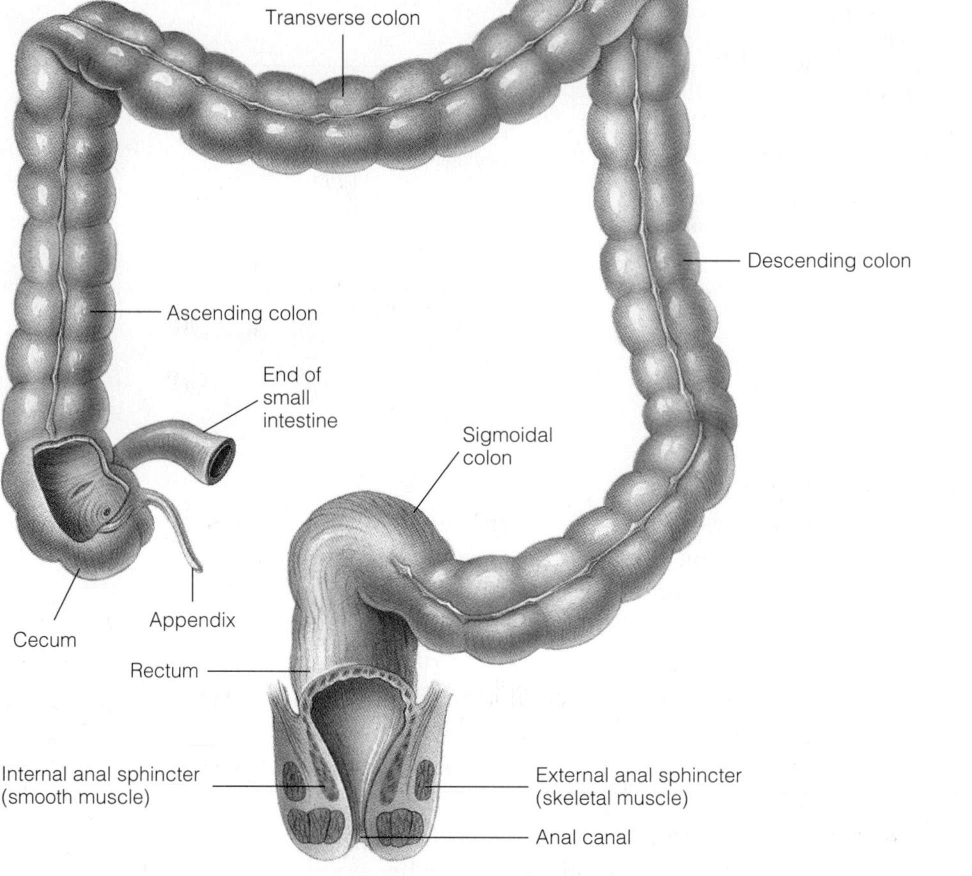

FIGURE 3-7 The Large Intestine This organ consists of four basic parts: the cecum, appendix, colon, and rectum.

The Large Intestine

After most of the nutrients have been absorbed, what's left of the food passes into the large intestine. The **large intestine** is about 1.5 meters (5 feet) long (**FIGURE 3-7**).

In the large intestine, much of the water, sodium, and potassium ions in the waste is absorbed, passing back into the blood stream. After water and salt have been removed, the contents of the large intestine are known as the *feces*. The feces consist primarily of undigested food and materials that cannot be digested, such as cellulose. It also contains lots of intestinal bacteria, which account for about one-third of the dry weight of the feces.

The feces are propelled along the large intestine by contractions of smooth muscle in the walls of the organ. The fecal matter accumulates in the rectum, the last part of the large intestine. This causes the rectum to expand. This, in turn, stimulates nerve receptors in the wall of the rectum. Nerve impulses from the receptors in the rectum travel via nerves to the spinal cord. In babies and very young children, the incoming nerve impulses stimulate nerve cells that supply the smooth muscle in the wall of the rectum. Impulses carried along these nerves cause the walls of the rectum to contract, expelling the feces.

In adults and older children, defecation does not occur until the **external anal sphincter** relaxes. This ring of muscles at the end of the anal canal is composed of skeletal muscle and is under conscious control in all individuals except babies. If the time and place are appropriate, the external anal sphincter is relaxed, and defecation can occur. Defecation is usually assisted by voluntary contractions of the abdominal muscles.

If circumstances are inappropriate, voluntary tightening of the external anal sphincter prevents defecation. When defecation is delayed, the muscle in the wall of the rectum relaxes, and the urge to defecate goes away—until the next movement of feces into the rectum occurs. If defecation is delayed too long, additional water may be removed from the feces, making them hard and dry and difficult to pass. This condition is called **constipation**.

Besides being uncomfortable, constipation may result in a dull headache, loss of appetite, and depression. Constipation can also be caused by decreases in contractions of the large intestine, which occur with age, emotional stress, or low-fiber, high meat diets and other factors.

Constipation can also result in serious problems. Hardened fecal material, for instance, may become lodged in the appendix, a small wormlike organ attached to the very first part of the large intestine. This, in turn, may lead to inflammation of the organ, a condition called **appendicitis** (ah-PEND-eh-SITE-iss). When this occurs, the appendix becomes swollen and filled with pus and must be surgically removed to prevent the otherwise useless organ from bursting and spilling its contents into the abdominal cavity. Fecal matter leaking into the abdominal cavity introduces billions of bacteria and can result in a deadly infection.

Controlling Digestion

Digestion is a complex process controlled by nerves and hormones. This section discusses some of the key events involved in the control of digestion.

Digestion begins in the oral cavity, as noted earlier. The sight, smell, taste, and sometimes even the thought of food stimulates the release of saliva. Chewing has a similar effect. The secretion of saliva via these routes is controlled by the nervous system and is largely a reflex response.

Besides activating the release of saliva, the stimuli listed above also cause the brain to send nerve impulses to the stomach. These nerve impulses initiate the secretion of stomach acid (HCl) and protein-digesting enzymes from the glands in the stomach's lining. The most potent stimulus for the release of these substances, however, is the presence of protein in the stomach.

Food entering the small intestine stimulates the release of two hormones by the small intestine. They circulate in the blood and enter the pancreas. Here, they stimulate the release of the pancreatic juice, containing enzymes and sodium bicarbonate.

The digestion and absorption of food requires the actions of many organs belonging to the digestive system. The processes involved are intricately coordinated to ensure that food is broken down into molecules that the cells of our bodies can use.

The Circulatory System

The human circulatory system is a marvel of structure and function. It consists of a powerful muscular pump, the heart, which beats approximately 100,000 times per day (FIGURE 4-1). If you had a dollar for every heartbeat, you'd be a millionaire in 10 days. The heart pumps blood through an extensive, branching network of blood vessels that, placed end to end, covers 50,000 miles.

Blood circulating throughout the body carries oxygen from the lungs and nutrients absorbed by the digestive system to the cells, tissues, and organs of the body. In addition, the circulatory system picks up waste from cells in the body's tissues and transports it to organs that get rid of it for us, primarily the kidneys and liver. It also helps to distribute body heat.

The Heart

The heart is located in the chest between the lungs. The walls of the heart, shown in FIGURE 4-2, are composed of three layers, the pericardium, the myocardium, and the endocardium. The pericardium (pear-ah-CARD-ee-um) is a thin, closed sac that surrounds the heart and the bases of large vessels that enter and leave this organ. It is filled with a clear, slippery, watery fluid that reduces friction produced when the heart contracts. The middle layer, the myocardium (my-oh-CARD-ee-um), is the thickest part of the wall. It is composed chiefly of cardiac muscle cells. The inner layer, the endocardium (end-oh-CARD-ee-um), is the endothelial layer.

The heart pumps blood through two distinct but connected circuits, shown in FIGURE 4-1B. They are the pulmonary circuit, which carries blood to and from the lungs, and the systemic circuit, which transports blood to and from the rest of the body.

The pulmonary circuit supplies blood with oxygen and gets rid of carbon dioxide, a waste product of cellular energy production. After blood is oxygenated in the pulmonary circuit, it is distributed throughout the body in the systemic circuit.

As shown in Figure 4-1b, the heart consists of four hollow chambers—two on the right

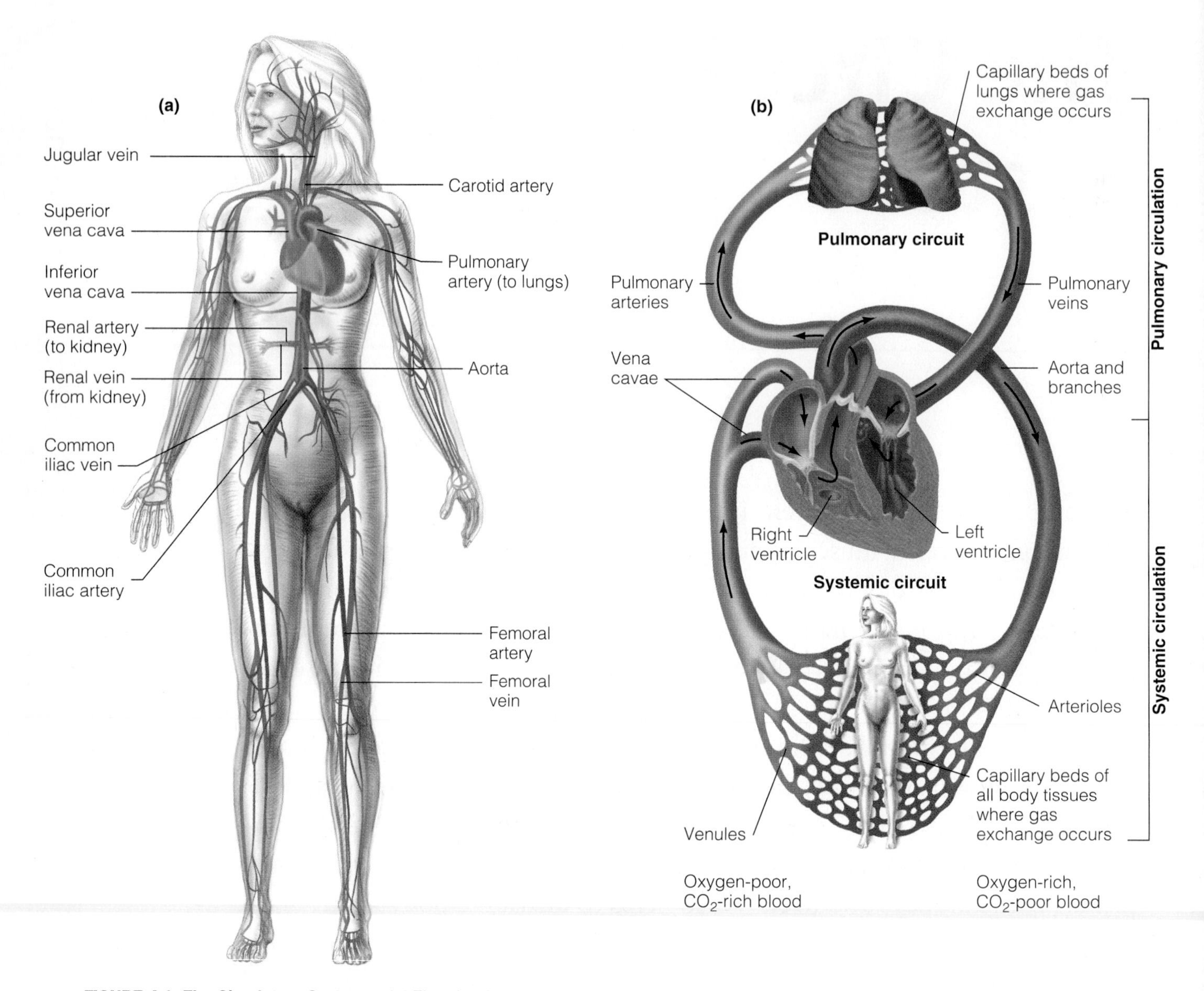

FIGURE 4-1 The Circulatory System *(a)* The circulatory system consists of a series of vessels that transport blood to and from the heart, the pump. *(b)* The circulatory system has two major circuits, the pulmonary circuit, which transports blood to and from the lungs, and the systemic circuit, which transports blood to and from the body (excluding the lungs).

side of the heart and two on the left. The right side of the heart pumps blood through the pulmonary circuit. The left side of the heart pumps blood through the systemic circuit.

FIGURE 4-2 illustrates the course that blood takes through the heart. Let's start with blood returning from the body in the systemic circulation. Drawn in blue, blood low in oxygen (and rich in carbon dioxide) enters the right side of the heart from two large veins draining the body, the superior and inferior vena cavae (VEEN-ah CAVE-ee). These veins empty directly into the right atrium (A-tree-um), the uppermost chamber of the heart. The blood is then pumped from here into the right ventricle (VEN-trick-el), the lower chamber on the right side. When the right ventricle is full, the muscles in its wall contract, forcing blood into the pulmonary arteries, which lead to the lungs.

In the lungs, this blood is oxygenated, then returned to the heart via the pulmonary veins. The pulmonary veins, in turn, empty directly into the left atrium, the upper chamber on the left side of the heart. It's the first part of the systemic circuit.

Next, the oxygen-rich blood is pumped to the left ventricle. When it's full, the left ventricle's thick, muscular walls contract and propel the blood into

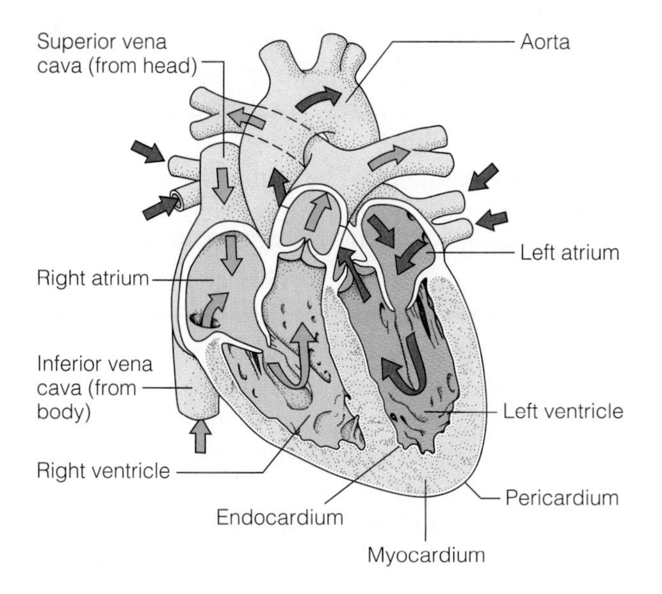

FIGURE 4-2 Blood Flow Through the Heart Deoxygenated (carbon-dioxide-enriched) blood (blue arrows) flows into the right atrium from the systemic circulation and is pumped into the right ventricle. The blood is then pumped from the right ventricle into the pulmonary artery, which delivers it to the lungs. In the lungs, the blood releases its carbon dioxide and absorbs oxygen. Reoxygenated blood (red arrows) is returned to the left atrium, then flows into the left ventricle, which pumps it to the rest of the body through the systemic circuit.

the aorta (a-OR-tah). The aorta is the largest artery in the body. It carries the oxygenated blood away from the heart, delivering it to the body.

The flow of blood just described presents a slightly misleading view of the way the heart works. As shown in **FIGURE 4-3**, both atria actually fill and contract simultaneously, delivering blood to their respective ventricles. The right and left ventricles also fill simultaneously, and when both ventricles are full, they too contract in unison. Blood is then pumped into the systemic and pulmonary circulations. The coordinated contraction of heart muscle is brought about by an internal timing device, or pacemaker (described later).

Heart Valves. The human heart contains four valves that control the direction of blood flow, ensuring a steady flow (**FIGURE 4-4**). The valves between the atria and ventricles are known as *atrioventricular valves* (AH-tree-oh-ven-TRICK-u-ler). Each valve consists of two or three flaps of tissue anchored to the inner walls of the ventricles by slender cords (Figure 4-4). Between the right and left ventricles, and the arteries into which they pump blood (pulmonary artery and aorta, respectively) are two more valves.

The valves of the heart are all one-way valves. That is, they permit the one-way flow of blood. When the ventricles contract, for example, blood forces the valves in the large arteries connecting to them to open. Blood flows out of the ventricles into the arteries. The backflow of blood causes the valve to shut, preventing blood from draining back into the ventricles.

Heart Sounds. When a doctor listens to your heart, she is listening to sounds of the heart valves closing. The noises she hears are called the heart sounds and are often described as "LUB-dupp." The first heart sound (LUB) results from the closure of the atrioventricular valves. It is longer and louder than the second heart sound (dupp), produced when

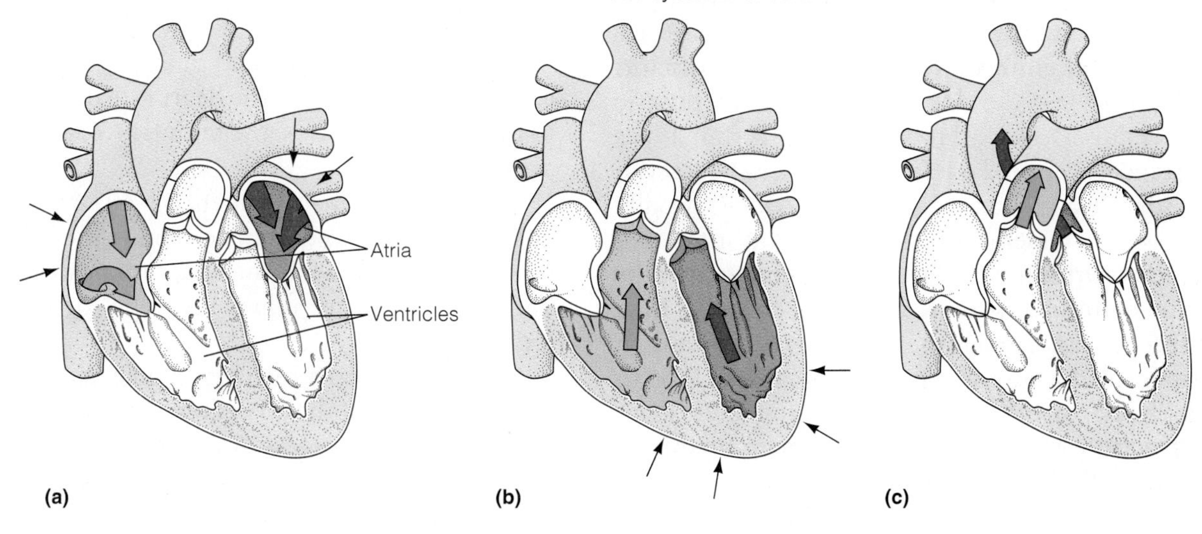

FIGURE 4-3 Blood Flow Through the Heart *(a)* Blood enters both atria simultaneously from the systemic and pulmonary circuits. When full, the atria pump their blood into the ventricles. *(b)* When the ventricles are full, they contract simultaneously, *(c)* delivering the blood to the pulmonary and systemic circuits.

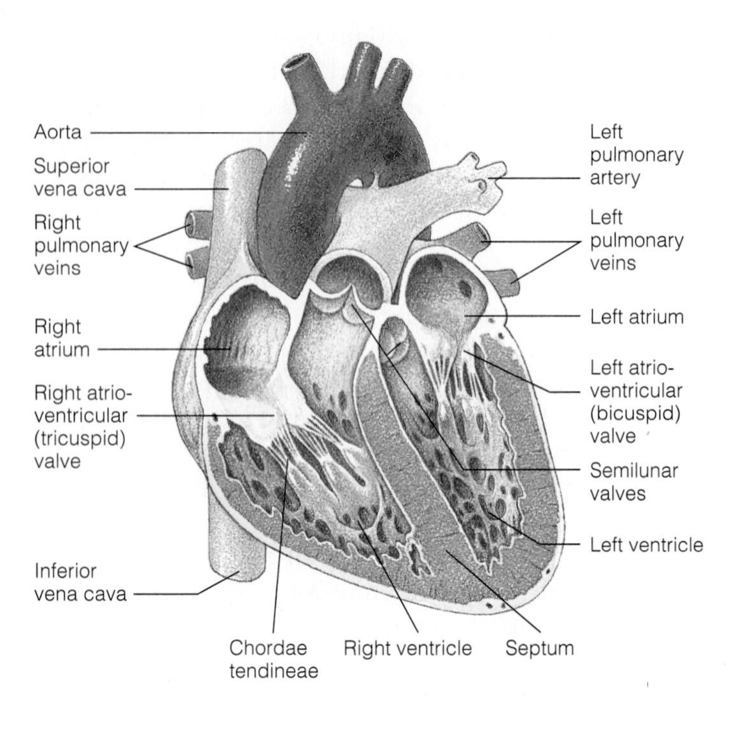

FIGURE 4-4 Heart Valves A cross section of the heart showing the four chambers and the location of the major vessels and valves.

the valves in the connecting arteries shut. Because the valves do not close at the same time, careful placement of the stethoscope allows a doctor to listen to each valve individually to determine whether it is working properly.

For most of us, our heart valves function flawlessly throughout life. However, in some individuals, diseases alter the valves. This, in turn, may decrease the efficiency of the heart and the circulation of blood. Rheumatic (RUE-mat-tick) fever, for example, is caused by a bacterial infection and affects many parts of the body, including the heart. Although it is relatively rare in wealthier countries, rheumatic fever is still a significant problem in less developed countries. Rheumatic fever begins as a sore throat caused by certain types of streptococcus (STREP-toe-COCK-iss) bacteria. The sore throat—known as *strep throat*—is usually followed by general illness. During this infection, the body forms antibodies (proteins made by cells of the immune system) to the bacteria. These antibodies can damage the heart valves, preventing them from closing completely. This causes blood to leak back into the atria and ventricles after contraction and results in a distinct "sloshing" sound, called a *heart murmur*. This condition reduces the efficiency of the heart and causes the organ to work harder to make up for the inefficient pumping. In severe cases, increased activity can result in heart failure.

The Heart's Pacemaker. The human heart functions at different rates under different conditions. At rest, it generally beats slowly. When one is working hard, it beats much faster. This change in heart rate helps the body adjust for differences in oxygen requirements by cells and tissues.

Heart rate is controlled by a number of mechanisms. One of the most important is an internal pacemaker, the sinoatrial (SA) node (SIGN-oh-A-tree-ill noad). Located in the wall of the right atrium, the SA node is composed of a clump of specialized cardiac muscle cells. These cells produce a tiny electric impulse, like those produced by nerve cells. This impulse spreads rapidly from the node to the rest of heart muscle, first in the atria and then into the ventricles. It therefore controls the contraction of the entire heart. Without it the muscle cells would contract independently, and the heart would be ineffective.

The electrical impulse created by the SA node is slowed briefly as it passes from the atria to the ventricles. This gives the blood-filled atria time to contract and empty their contents into the ventricles. This delay also provides the ventricles plenty of time to fill before they are stimulated to contract.

The SA node of the human heart produces a steady rhythm of about 100 beats per minute, but this is much too fast for most human activities. To bring the heart rate in line with body demand, the SA node must be slowed down by impulses transmitted by nerves from a heart control center in the brain. These impulses slow the heart to about 70 beats per minute when you are at rest. During exercise or stress, when the heart rate must increase to meet body demands, the impulses from the brain are reduced.

Other nerves also influence heart rate. These nerves carry impulses that accelerate the heart rate even further, allowing the heart to attain rates of 180 beats or more when the cells' demand for oxygen is great.

Several hormones also play a role in controlling heart rate. One of these is adrenalin. This hormone is secreted during stress or exercise by the adrenal glands located on top of the kidneys. Adrenalin accelerates the heart, increasing the flow of blood through the body.

Electrical Activity in the Heart. Electrical changes in the cardiac muscle cell occur when the muscles of the atria and ventricles contract. These changes can be detected by doctors by placing electrodes—small metal plates connected to wires and a voltage meter—on the chest. The resulting reading on the voltage meter is called an electrocardiogram (ECG or EKG). Diseases of the heart may result in noticeable changes in the EKG. As a result, an EKG is often a valuable diagnostic tool for heart doctors, cardiologists.

Heart Attacks: Causes, Cures, and Treatments

Many Americans die each year from heart attacks. Those who survive must often undergo surgery and make dramatic changes in their diets to prevent another attack.

The most common type of heart attack is known as a *myocardial infarction* (my-oh-CARD-ee-al in-FARK-shun) or MI. MIs are caused by blood clots that block one or more of the arteries that supply the heart, but usually only if they are narrowed by plaque (discussed in Module 2). A blood clot lodged in a heart artery restricts the flow of blood to the heart muscle, cutting off the supply of oxygen and nutrients. The lack of oxygen can damage and kill the heart muscle cells. The damaged region is called an *infarct* (in-FARKT)—hence the name *myocardial infarction*.

The formation of plaque results from a combination of factors: smoking, poor diet, lack of exercise, heredity, stress, and others. Note, however, narrowing of a coronary artery by plaque will usually not block the flow of blood enough to cause a heart attack unless it is quite severe—around 80% or 90% blockage. Less severe narrowing, however, can cause blood clots to form in the vessel at the site of narrowing. Blood clots can block the artery, causing a heart attack. Clots can also form in other parts of the body and may travel in the bloodstream. When they reach the narrowing in a heart artery, they can become jammed in the artery, blocking blood flow. If the size of the area that is damaged by lack of oxygen is small, a heart attack is usually not life-threatening. If the damage is great, a heart attack can prove fatal.

Getting help quickly lessens the chances of severe damage to the heart. Giving a person an aspirin during a heart attack also reduces the damage because aspirin reduces clotting.

Heart attacks can occur quite suddenly, without warning, or may be preceded by several weeks by a type of pain called *angina* (an-GINE-ah). Angina occurs when the supply of oxygen to the heart muscle is reduced. The pain appears in the center of the chest and can spread to a person's throat, upper jaw, back, and arms (usually just the left one). Angina is a dull, heavy, constricting pain that appears when an individual is active, then disappears when he or she ceases the activity.

Angina may also be caused by stress and exposure to carbon monoxide, a pollutant that reduces the oxygen-carrying capacity of the blood. Angina begins to show up in men at age 30 and is nearly always caused by coronary artery disease. In women, angina tends to occur at a much later age.

Another type of heart attack occurs when the SA node loses control of the heart muscle. When this occurs, the cardiac muscle cells beat independently. The result is that the heart reduces, even stops, pumping blood. If the heart stops beating altogether, the condition is known as *cardiac arrest*.

Physicians treat this type of heart attack by applying a strong electrical current to the chest. It is often sufficient to restore normal electrical activity and heartbeat. A normal heartbeat can also be restored by cardiopulmonary resuscitation (CPR), in which the heart is "massaged" externally by applying pressure to the breastbone.

Prevention Is the Best Cure. Heart disease is not inevitable. It can be avoided by a healthy diet, exercise, and stress management. A half a tablet of aspirin a day taken over long periods can also help reduce an individual's risk of a heart attack. Quitting smoking also greatly cuts down on one's risk of a heart attack.

Heart attacks are also treated by using one of several blood-clot-dissolving agents. When given within a few hours of the onset of a heart attack, they can greatly reduce the damage to heart muscle and accelerate a patient's recovery.

In mild heart attacks, physicians can open clogged blood vessels by inserting a small device called a **catheter** into the heart artery. A tiny balloon attached to its tip is then inflated. After chemical

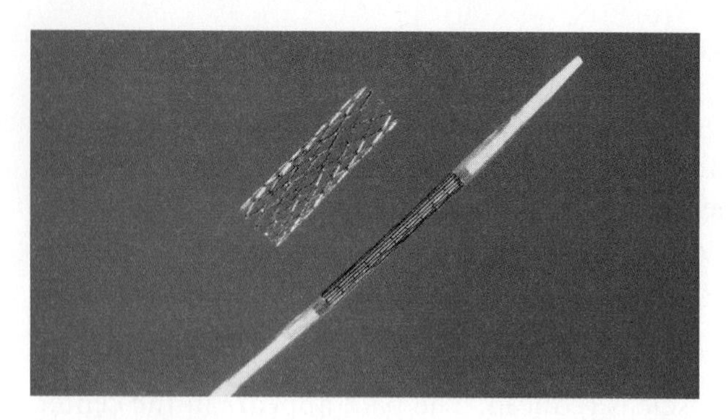

FIGURE 4-5 **Stent**

clot dissolvers are injected, the balloon is inflated. This forces the artery open and loosens the plaque from the wall. This procedure is called *balloon angioplasty* (AN-gee-oh-PLAST-ee). Scientists are also experimenting with lasers that burn away plaque in artery walls. Unfortunately, as in other techniques, cholesterol builds up again in the walls of arteries within a few months.

To avoid plaque clogging an artery opened by balloon angioplasty, surgeons often insert a tiny device called a *stent* into the artery after balloon angioplasty (FIGURE 4-5). After the balloon is removed, the stent holds the artery open.

In severe cases, the coronary arteries may be completely blocked by plaque. To restore blood flow to the heart, surgeons often perform coronary bypass surgery. In this procedure, surgeons transplant small segments of veins from the leg into the heart (FIGURE 4-6). The veins are connected to the artery on either side of the clogged area so blood can bypass it. Unfortunately, the veins often fill fairly quickly with plaque.

The Blood Vessels

While the heart serves as a central pump, it is the blood vessels that transport blood throughout the body. The blood vessels form an extensive network of ducts in the body. Three types of blood vessels are present: arteries, capillaries, and veins.

Arteries carry blood away from the heart. As they travel through the body, arteries branch off from time to time to supply various organs. Within each organ, arteries branch again, forming smaller and smaller vessels. The smallest of all arteries is the **arteriole** (are-TEAR-ee-ol).

As shown in FIGURE 4-7, arterioles empty into capillaries (CAP-ill-air-ees), tiny, thin-walled vessels that permit wastes and nutrients to pass through with relative ease. **Capillaries** form extensive, branching networks in body tissues, referred to as *capillary beds*. Capillaries have very thin walls which permit water and various nutrients to pass from the blood into the surrounding tissues and allow waste to move in the opposite direction.

Blood flows out of the capillaries into the smallest of all veins, the venules. Venules, in turn, converge to form small veins, which unite with other small veins, forming larger veins. Blood in veins flows toward the heart.

Blood Pressure. As blood is pumped throughout the circulatory system, it creates a force against the walls of the blood vessels. Created by the pumping of

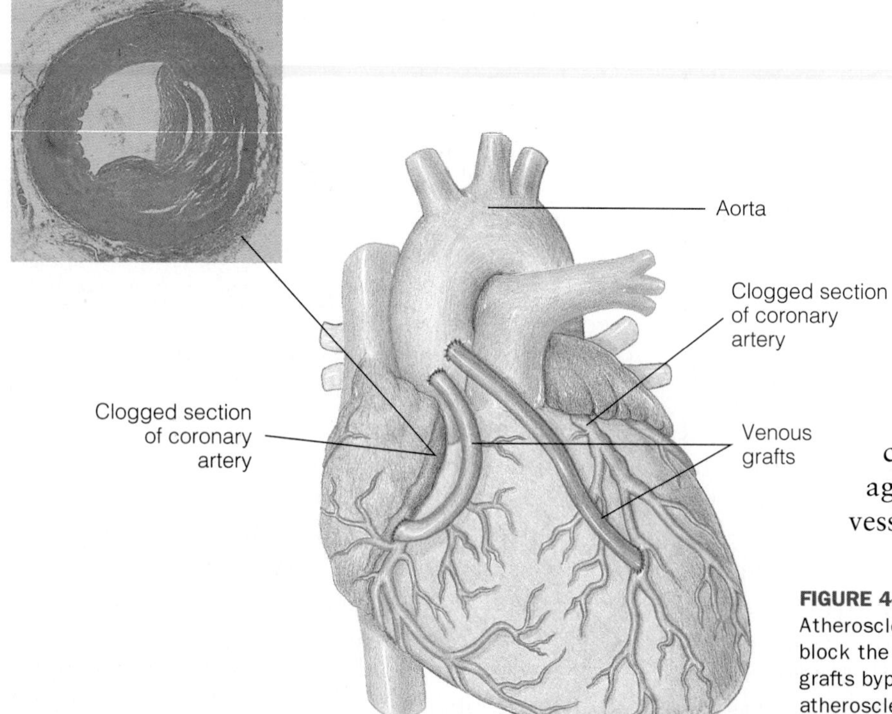

FIGURE 4-6 **Coronary Bypass Surgery** *(a)* Atherosclerotic plaque in coronary arteries can block the flow of blood to heart muscle. *(b)* Venous grafts bypass coronary arteries blocked by atherosclerotic plaque.

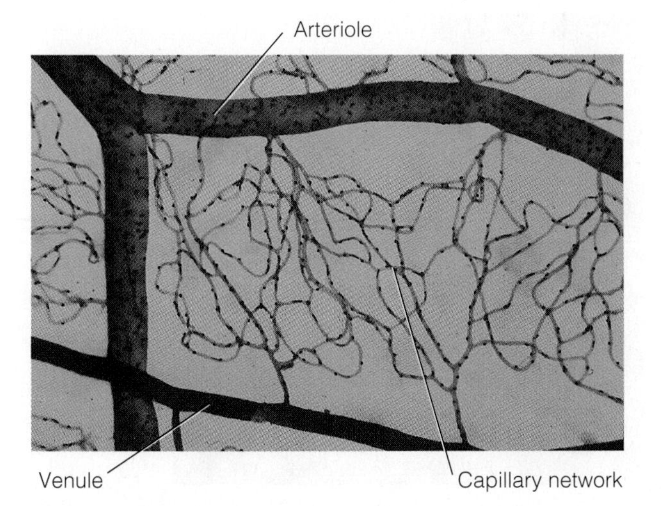

FIGURE 4-7 Capillary Network A network of capillaries between the arteriole and the venule delivers blood to the cells of body tissues (not shown).

the heart, this force is known as the *blood pressure.* Blood pressure varies throughout the cardiovascular system, being the highest in the arteries and the lowest in the capillaries and veins.

Like many other physical conditions in the human body, blood pressure varies from time to time. For example, it changes in relation to stress levels. When someone makes you angry or stresses you out, they really are raising your blood pressure! Blood pressure also increases with age and when arteries become hardened by plaque. High blood pressure is an indication of cardiovascular disease.

Blood pressure is measured with a device called a *blood pressure cuff* (**FIGURE 4-8**). The blood pressure cuff is wrapped around the upper arm. A stethoscope is placed over an artery just below the cuff. Air is pumped into the cuff until the pressure stops the flow of blood through the artery. The pressure in the cuff is then gradually reduced as air is released. When the blood pressure in the artery exceeds the pressure of the cuff, the blood starts flowing through the vessel once again. This point represents the systolic pressure (sis-STOL-ick), the pressure at the moment the ventricles contract. Systolic pressure is the higher of the two numbers in a blood pressure reading (120/70, for example). The pressure at the moment the heart relaxes to let the ventricles fill again is the diastolic pressure (DIE-ah-STOL-ick) and is the lower of the two readings. It is determined by continuing to release air from the cuff until no arterial pulsation can be heard. At this point, blood is flowing continuously through the artery.

A typical reading for a young, healthy adult is about 120/70, although readings vary considerably from one person to the next. Thus, what is normal for one person may be abnormal for another. As a person ages, blood pressure tends to rise. Thus, a healthy 65-year-old might have a blood pressure reading of 140/90.

Hypertension is a prolonged elevation in blood pressure. Like other diseases of the cardiovascular system, it has many causes, including obesity. The problem with hypertension is that blood pressure increases gradually over time. A person may feel fine for years. Symptoms, such as headaches, rapid, forceful beating of the heart (palpitations), and a general feeling of ill health usually occur only when blood pressure is dangerously high. Consequently, early detection and treatment are essential to prevent serious problems, including heart attacks.

How the Capillaries Function. The heart, arteries, and veins form an elaborate system that transports blood to and from the capillaries. It is here that

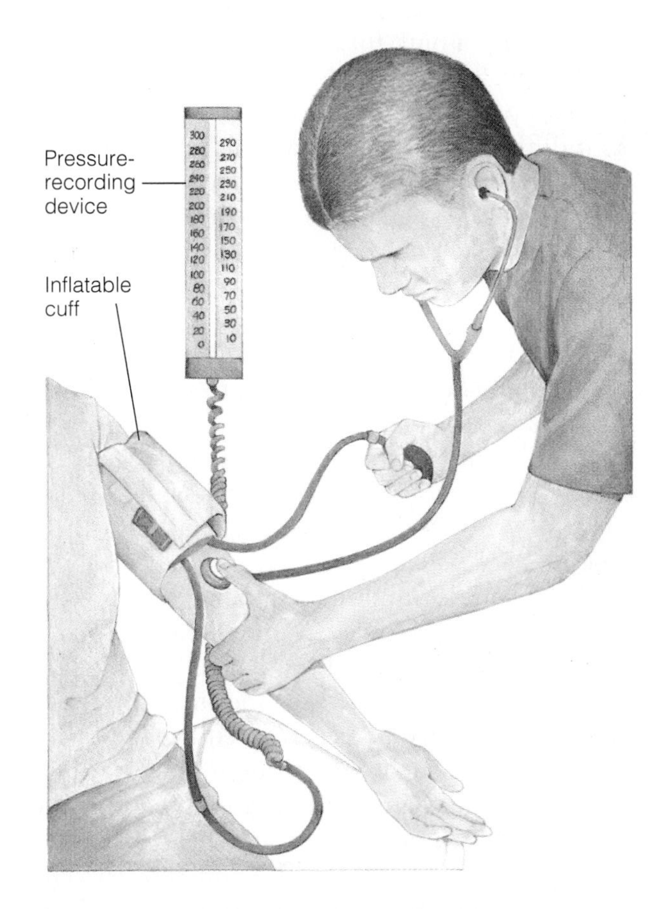

FIGURE 4-8 Blood Pressure Reading A sphygmomanometer (blood pressure cuff) is used to determine blood pressure.

wastes and nutrients are exchanged between the cells of the body and the blood.

As blood flows into a capillary bed, nutrients, gases, water, and hormones carried in the blood immediately begin to diffuse out of the tiny vessels. Meanwhile, water-dissolved wastes in the tissues, such as carbon dioxide, begin to diffuse inward.

Rapid exchange is possible because blood pressure is very low and thus blood slows down considerably as it passes through the capillaries. Rapid exchange is also possible because the walls of capillaries are very thin, consisting of only a single layer of flattened cell. These cells permit dissolved substances to pass through them with ease.

The constriction and dilation of the arterioles that "feed" the capillaries help to regulate blood flow through the capillaries. They also help to regulate body temperature. On a cold winter day, for example, the arterioles close down, restricting blood flow through the capillaries. This conserves body heat. On a warm day, the flow of blood through the skin increases. This releases heat and often creates a pink flush.

How Veins Function. Blood leaves the capillary beds stripped of its nutrients and loaded with waste. As it drains from the capillaries, the blood enters the smallest of all veins. They unite to form larger veins. Eventually, all of the blood returning to the heart in the systemic circuit enters the superior or inferior vena cavae that drain into the right atrium of the heart (Figure 4-2). These vessels drain the upper and lower parts of the body, respectively.

As noted earlier, blood pressure in the veins is low, and veins have relatively thin walls. Because the veins' walls are so thin, obstructions can cause blood to pool in them, creating bluish bulges called *varicose veins* that can be quite painful (VEAR-uh-cose). See **FIGURE 4-9**.

Some people inherit a tendency to develop varicose veins. However, most cases can be attributed to various factors that reduce the flow of blood back to the heart: abdominal tumors, pregnancy, obesity, and even sedentary life-styles.

Varicose veins may also form in the wall of the anal canal. The veins in this region are known as the *internal hemorrhoidal veins* (hem-eh-ROID-il). A swelling of the internal hemorrhoidal veins results in a condition known as *hemorrhoids* (hem-eh-ROIDS). Because the internal hemorrhoidal veins are supplied by numerous pain fibers, this condition can be quite painful.

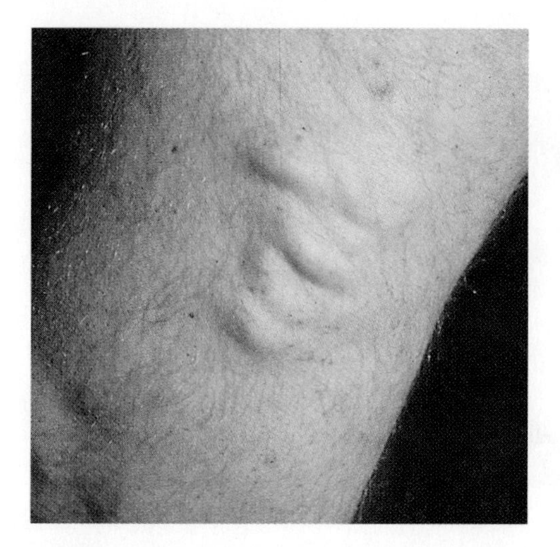

FIGURE 4-9 Any restriction of venous blood flow to the heart causes veins to balloon out, creating bulges commonly known as *varicose veins.*

Blood flows through arteries because of blood pressure, but in veins it flows because of other forces. In veins located above the heart, gravity draws blood down to the heart. In veins below the heart, blood is propelled by the movement of body parts. As you walk, for example, the contraction of muscles forces the blood upward, causing it to move against the force of gravity.

Valves in the veins also help in this process. Valves are flaps of tissue that span the veins and prevent the backflow of blood (**FIGURE 4-10**). Just as in the valves of the heart, blood pressure, however slight, pushes the flaps open. This allows the blood to move forward. As the blood fills the segment of the vein in front of the valve, it pushes back on the valve flaps and forces them shut. Blood is then propelled through the next set of valves and so on and so on until it empties into the vena cavae.

The Lymphatic System

The lymphatic system is an extensive network of vessels and glands (**FIGURE 4-11**). It is functionally related to the circulatory system and the immune system. This module examines the circulatory role of the lymphatic system.

The cells of the body are bathed in a liquid called *tissue fluid*. It provides a medium through which nutrients, gases, and wastes can diffuse between the capillaries and the cells.

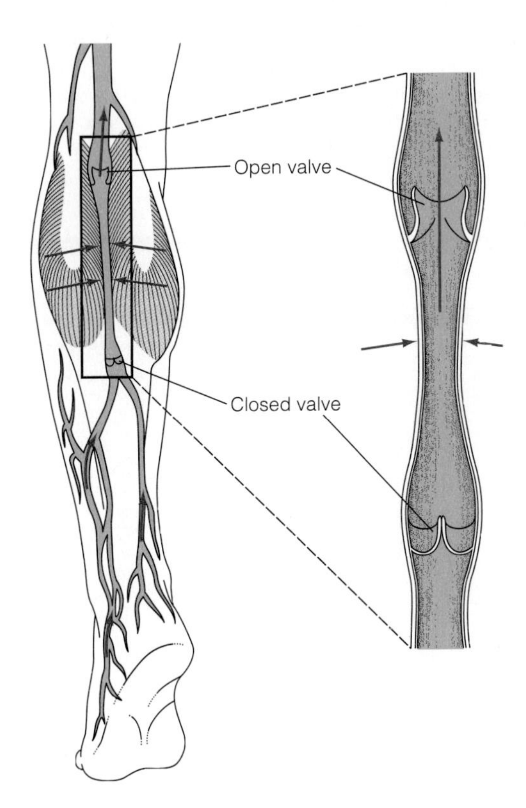

FIGURE 4-10 **Valves in Veins** The slight blood pressure in the veins and the contraction of skeletal muscles propel the blood along the veins back toward the heart. The one-way valves stop the blood from flowing backward.

Tissue fluid is replenished by water from capillaries. The flow of water out of the capillaries, however, normally exceeds the return flow by about 3 liters (nearly 3 quarts) per day. What happens to this excess? It is picked up by small lymph vessels, called *lymph capillaries*, located in all of the tissues and organs of the body. These vessels have thin walls through which water and other substances easily pass.

Lymph drains from the capillaries into larger ducts. These vessels, in turn, merge with others, creating larger and larger ducts. They eventually empty into the large veins at the base of the neck.

Lymph moves through the vessels of the lymphatic system in much the same way that blood is transported in veins. In the upper parts of the body, it flows by gravity. In regions below the heart, it is propelled largely by muscle contraction. Lymph flow is also assisted by valves.

The lymphatic system also consists of several lymphatic organs. Because they function in protecting the body, we will discuss them in Module 6.

Normally, lymph is removed from tissues at a rate equal to its production. In some instances, however, lymph production exceeds its uptake. A burn, for example, may cause extensive damage to capillaries, which causes them to release excess water into tissues. This flood overwhelms the lymph vessels, causing a buildup of fluid in tissues.

The heart and blood vessels are essential to all cells and organs of the body—and to our health. Nutritious food, exercise, rest, stress control, and a healthy life-style are vital to ensuring that our heart and blood vessels remain healthy so we can live long, productive lives.

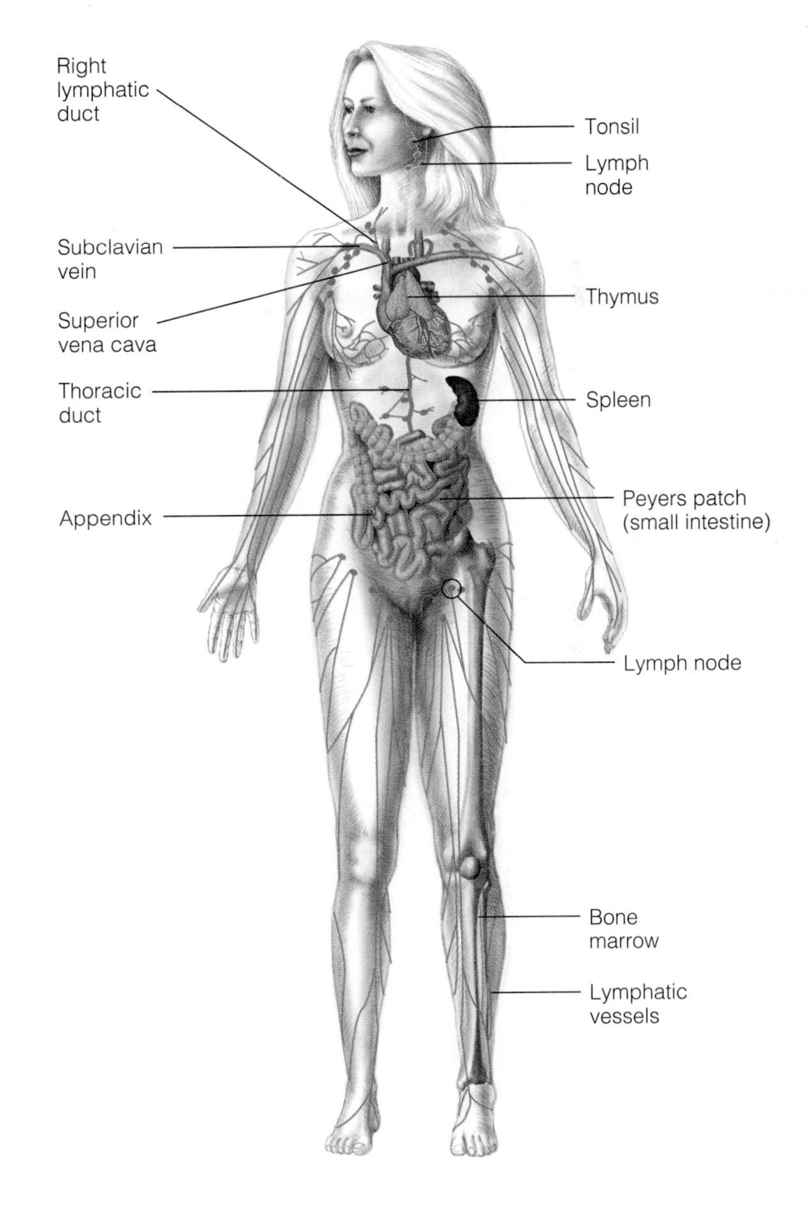

FIGURE 4-11 **The Lymphatic System** The lymphatic system consists of vessels that transport lymph, excess tissue fluid, back to the circulatory system.

The Blood

The circulatory system, described in the previous module, has one purpose: to transport blood throughout the body. The blood, in turn, supplies nutrients to body cells and removes wastes. But blood also performs a number of other vital functions, which you will learn about in this module.

Composition of Blood

Blood consists of a fluid known as *plasma*. **Plasma** is mostly water containing numerous dissolved substances such as sodium, glucose, and blood proteins. Blood also contains white blood cells, red blood cells, and platelets, described shortly.

Blood accounts for about 8% of our total body weight. A man weighing 70 kilograms (150 pounds), has about 5 to 6 liters (1.3 to 1.5 gallons) of blood. On average, women have about a liter less.

Plasma makes up about 55% of the blood volume. White blood cells, red blood cells, and platelets make up about 45% (**FIGURE 5-1**). Most of the blood cells consist of red blood cells. However, the number of red blood cells varies in individuals living at different altitudes. In Denver, Colorado, a city located a mile above sea level, for example, more red blood cells are present to make up for the lower oxygen levels in the atmosphere. Blood cells and platelets therefore typically constitute about 50% of the blood volume. Because higher concentrations of red blood cells can give athletes a competitive advantage, many athletes train at higher altitudes. It is partly for this reason that the U.S. Summer Olympic team is headquartered in Colorado Springs, Colorado.

With this information in mind, let's take a deeper look at each of the main components of blood to learn more about what they do.

Plasma

Plasma performs many vital functions in the body. It transports macronutrients such as glucose, lipids, and amino acids and micronutrients such as sodium, chloride, and calcium from the digestive system to the cells of the body. Besides nourishing cells and removing wastes, the blood also helps to regulate body functions. It does this by carrying hormones, chemical messengers, from their site of production to their site of action. Plasma also contains proteins that form blood clots, protecting us from bleeding to death when we cut ourselves. Other plasma proteins, such as antibodies, discussed in the next module, are involved in providing protection from disease. The concentration of many of these substances is finely controlled to ensure that the body operates well and we remain in good health.

Red Blood Cells

The **red blood cell** is the most abundant cell in human blood. Human RBCs are highly flexible disks, concave on both sides. RBCs, as most readers probably already know,

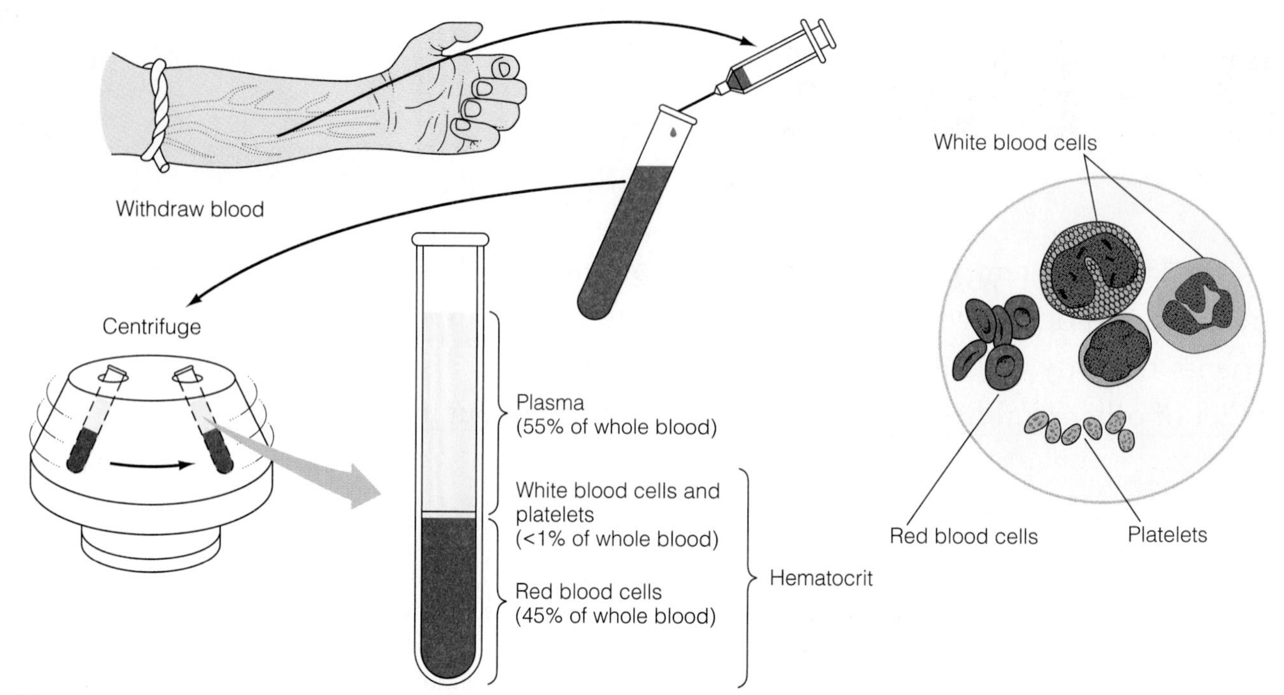

FIGURE 5-1 Blood Composition Blood removed from a person can be centrifuged to separate plasma from the cellular component. Red blood cells constitute about 45% of the blood volume, except at higher altitudes where they make up about 50% of the volume to compensate for the lower oxygen levels.

transport oxygen and, to a lesser degree, carbon dioxide in the blood (**FIGURE 5-2**).

Swept along in the bloodstream, RBCs travel throughout the circulatory system, each day traveling from arteries to capillaries to veins, then back again. As they pass through the capillaries, the RBCs bend and twist. This permits them to pass through the many miles of tiny capillaries without clogging them.

The flexibility of RBCs is important for our health and survival. Without it, RBCs could easily logjam in capillaries, blocking blood flow. But not everyone is so fortunate. Approximately one of every 500–1000 African Americans suffers from a disease called **sickle-cell anemia** (ah-NEEM-ee-ah). This disease is characterized by a decrease in the flexibility of RBCs. It results from a slight genetic mutation—an alteration in the hereditary material (DNA). RBCs in such individuals become sickle-shaped cells and inflexible when they encounter low levels of oxygen in capillaries (**FIGURE 5-3**). Sickle-shaped cells collect at branching points in capillary beds. This blocks blood flow, disrupting the supply of nutrients and oxygen to tissues and organs. The lack of oxygen causes considerable pain, but can also kill body cells. Blockages in the heart and brain often lead to heart attacks and serious brain damage. Many people who have sickle-cell anemia die in their late twenties and thirties; some die even earlier.

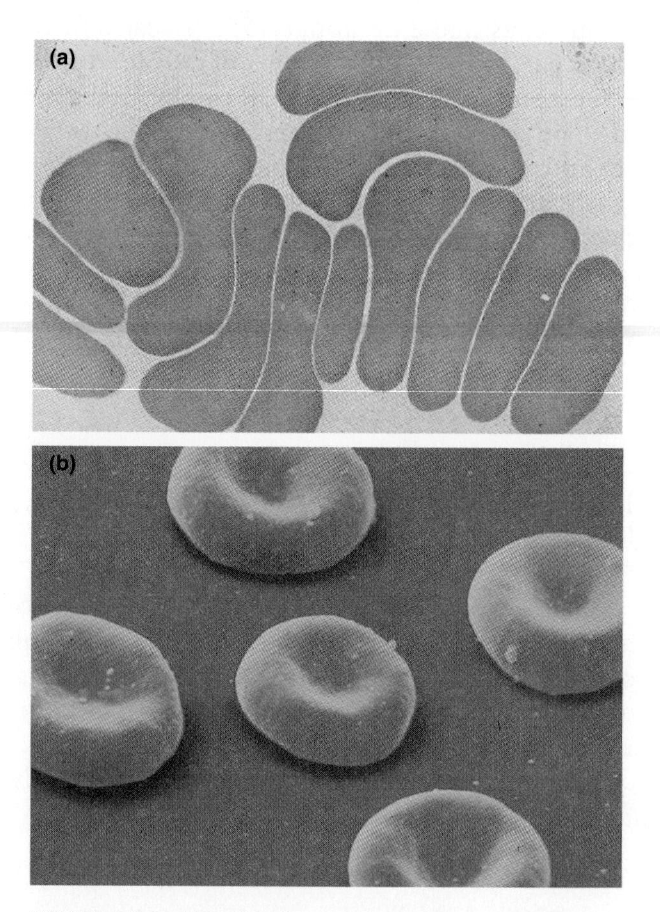

FIGURE 5-2 Red Blood Cells *(a)* Transmission electron micrograph of human RBCs showing their flexibility and their lack of organelles. *(b)* Scanning electron micrograph of human RBCs.

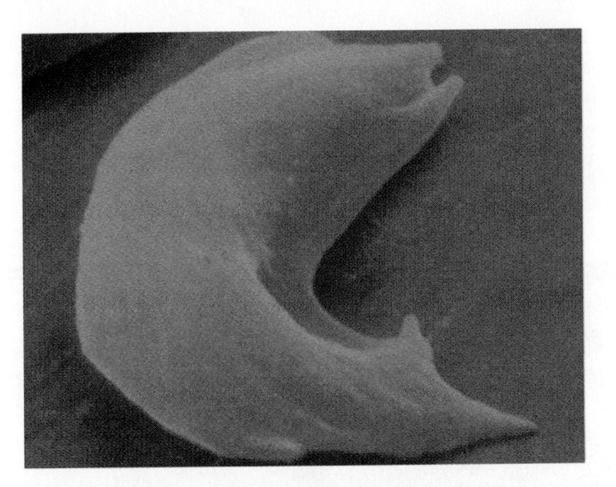

FIGURE 5-3 Sickle-Cell Anemia Scanning electron micrograph of a sickle cell.

On average, RBCs live about 120 days. At the end of their life span, the liver and spleen remove old RBCs from circulation. The iron contained in the hemoglobin, however, is recycled by these organs and used to produce new RBCs in the red bone marrow. The recycling of iron is not 100% efficient, however, so small amounts of iron must be ingested each day in the diet.

Human RBCs are highly specialized cells that lose their nuclei and organelles during cellular differentiation (FIGURE 5-2A). Because of this, RBCs cannot divide to replace themselves as they age. In humans, new RBCs are produced in **red bone marrow** found in certain bones of the body. Red bone marrow produces about 2 million RBCs per second! Because of this, the number of RBCs in the blood remains more or less constant over long periods. Maintaining a constant concentration of RBCs is essential to homeostasis. RBC production is stimulated by a hormone produced by the kidney. This hormone, known as **erythropoietin** (eh-RITH-row-po-EAT-in) or EPO, is also one of a dozen or more performance-enhancing chemical substances being used by athletes.

Because RBCs have no organelles, they mostly consist of a cell membrane enclosing large amounts of hemoglobin. **Hemoglobin** (HEME-oh-GLOBE-in) is a large protein found exclusively in the RBCs. Hemoglobin contains iron, which binds to oxygen. Carbon dioxide, a waste product of cellular respiration, also binds to hemoglobin, but to a much lesser degree.

Human health depends on the ability of blood to transport oxygen throughout the body. A reduction in the oxygen-carrying capacity of the blood is known as **anemia**. Anemia results in weakness and fatigue. Individuals are often pale and tend to faint or become short of breath easily.

Anemia may result from excessive bleeding which reduces the number of RBCs in the blood stream. It can also be caused by a tumor in the red marrow that reduces RBC production. Several infectious diseases (such as malaria) also decrease the RBC concentration in the blood. Anemia can also result from a reduction in the amount of hemoglobin in RBCs. This, in turn, may be caused by iron deficiency.

White Blood Cells

White blood cells are part of the body's protective mechanism, a system that combats harmful bacteria and viruses and helps us maintain homeostasis. White blood cells are produced in the bone marrow and circulate in the bloodstream.

The body contains five types of white blood cells. Two of them are responsible for engulfing harmful bacteria. When they arrive at the "scene" of an infection, they escape through the walls of the capillaries, then migrate into the infected tissue, gobbling up bacteria and debris.

Another type of white blood cells, known as the **lymphocyte**, produces antibodies. Antibodies are proteins that bind to and eliminate foreign organisms such as bacteria. Another type of lymphocyte attacks foreign cells such as parasites and tumor cells directly. You'll learn more about the lymphocytes in the next module.

Like red blood cells, WBCs are involved in homeostasis—that is, in maintaining internal constancy. Their numbers can increase greatly during infections and other diseases. Increases and decreases in various types of WBCs can be used to diagnose many diseases. For example, a dramatic increase in lymphocytes and lower abdominal pain are usually signs of appendicitis, an infection of the appendix, a small, useless organ attached to the large intestine.

Diseases Involving WBCs

Like many other cells in the body, WBCs can malfunction. Some white blood cells, for example, can divide uncontrollably in the bone marrow. In other words, they become cancerous. The cancerous WBCs then enter the bloodstream.

A cancer of WBCs is called **leukemia** (lew-KEEM-ee-ah). WBCs fill the bone marrow, crowding out the cells that produce RBCs and platelets. This results in a decline in RBC production, which leads to anemia. It also results in a reduction in platelet production. Because platelets are involved in blood clotting, leukemia reduces normal clotting and increases internal bleeding. In addition, WBCs produced in leukemia are often incapable of fighting infection. Leukemia patients typically die from infections and internal bleeding.

The most serious type of leukemia is acute leukemia, so named because it kills victims quickly. Children are the primary victims of this disease. Leukemia can be treated by irradiating the bone marrow and by administering drugs that halt uncontrollable cell division.

Another common disorder of the WBCs is **infectious mononucleosis** (MON-oh-NUKE-clee-OH-siss), commonly called mono. This disease is caused by a virus transmitted through saliva and may be spread by kissing or by sharing silverware, plates, and drinking glasses. Common complaints include fatigue, aches, sore throats, and low-grade fever. Rest and drinking plenty of liquids is required to recover.

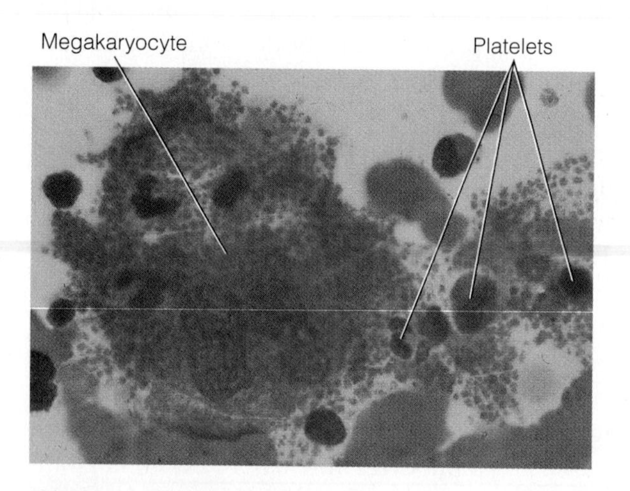

FIGURE 5-4 Megakaryocyte A light micrograph of a megakaryocyte, a large, multinucleated cell found in bone marrow; the megakaryocyte fragments, giving rise to platelets.

Platelets

Platelets are tiny cell fragments produced in the bone marrow from large platelet-producing cells (**FIGURE 5-4**). Like RBCs, platelets lack nuclei and organelles and, therefore, are not true cells. Also like RBCs, platelets are unable to divide. Platelets are coated by a layer of sticky material, which causes them to adhere to irregular surfaces, such as tears in blood vessels or plaque in arteries.

FIGURE 5-5 Blood Clot A scanning electron micrograph of a fibrin clot that has already trapped platelets and RBCs, plugging a leak in a vessel. The RBCs are red, and the fibrin network is blue.

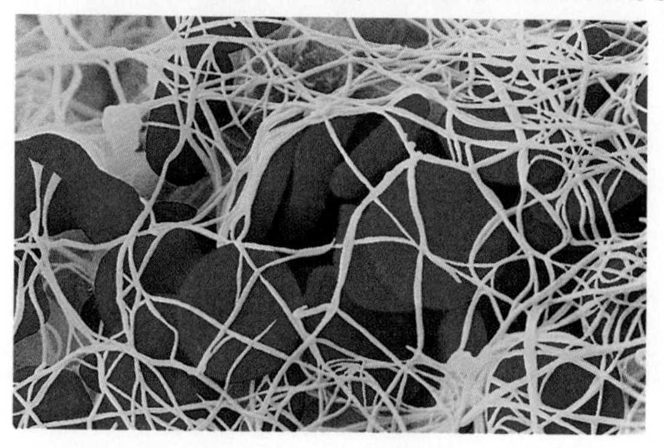

Blood clotting is actually a chain reaction involving several steps. When a blood vessel is torn, long, branching fibers made of a protein called **fibrin** collect at the damage point. They produce a weblike network in the wall of the damaged blood vessel (**FIGURE 5-5**). The fibrin web traps RBCs and platelets, forming a plug that stops the flow of blood to the tissue.

Blood clotting occurs fairly quickly. In most cases, a damaged blood vessel is sealed by a clot within 3–6 minutes of an injury. After the blood vessel is repaired, the clot is dissolved by a naturally occurring enzyme in the blood.

Blood clotting is a homeostatic mechanism vital to our survival, but in some individuals, blood clotting is impaired due to one of several problems such as an insufficient amount of platelets. (A reduced platelet count may result from leukemia.) In some individuals, the liver may not produce a sufficient amount of blood clotting factors, that is, substances required for normal blood clotting. The most common cause of this problem is a genetic defect known as **hemophilia** (he-moe-FEAL-ee-ah). Problems begin early in life, so even tiny cuts or bruises can bleed uncontrollably, threatening one's life. Because of repeated bleeding into the joints, victims suffer great pain and often become disabled; they often die at a young age. Fortunately, hemophiliacs can be treated by transfusions of blood-clotting factors.

The blood and circulatory system operate tirelessly transporting nutrients to body cells, removing wastes, sending hormonal messages to the cells of the body, repairing damaged vessels, and more. Although things can go wrong, for those who live a healthy lifestyle, this system performs admirably, keeping our bodies running smoothly and efficiently.

The Immune System

Floating in the air we breathe and water we drink are billions upon billions of microorganisms. Why don't we succumb to them? Most bacteria in our environment are harmless, incapable of causing disease. Those that are potentially harmful must evade our body's defenses, the topic of this module. Before we look at the ways the human body protects itself from such invasions, let's begin by looking at viruses and bacteria.

Viruses and Bacteria

Viruses consist of a strand of DNA or RNA surrounded by a protein coat, although some viruses posses an additional protective layer known as the *envelope* (**FIGURE 6-1**). On their own, viruses are pretty harmless. They are incapable of reproducing and are therefore not even considered to be living things.

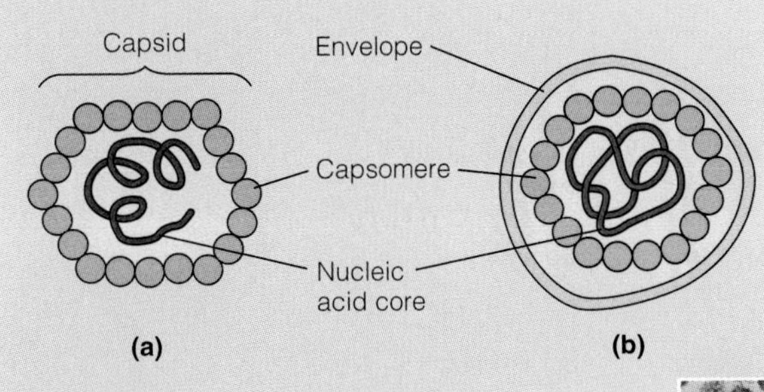

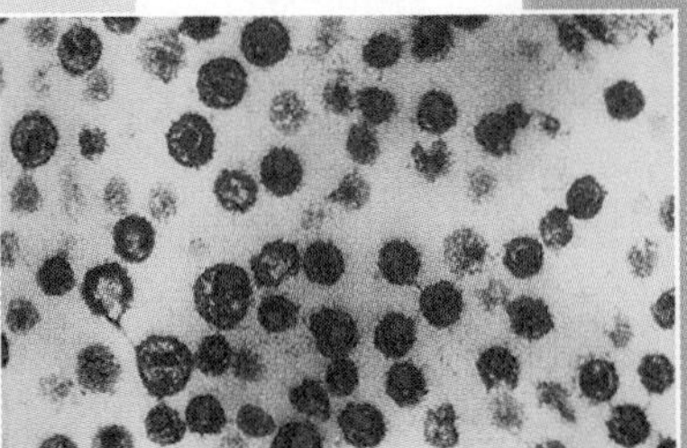

FIGURE 6-1 General Structure of a Virus *(a)* The virus consists of a nucleic acid core of either RNA or DNA. Surrounding the viral core is a layer of protein known as the capsid. Each protein molecule in the capsid is known as a capsomere. *(b)* Some viruses have an additional protective coat known as the envelope. *(c)* Electron micrograph of the Human Immunodeficiency Virus (HIV).

Viruses cause problems when they invade the body. Once in the body, they invade cells, taking over. Because of this, some scientists liken viruses to cellular pirates. Inside cells, viruses commandeer the cell's resources into making more viruses. In very short time, infected cells produce hundreds of thousands of new viruses. They are then released into the bloodstream and can therefore spread to other cells, infecting them.

Viruses most often enter the body through the respiratory and digestive systems. However, other avenues of entry are also possible—for example, the reproductive system.

Fortunately for us, the immune system kills many viruses, but usually not until we've undergone considerable suffering. Influenza viruses, which cause the flu, cause fevers, chills, headaches, and muscle pains that can last up to two weeks. During this time, the immune system mounts an attack, eliminating the virus in 10–14 days. Colds are also caused by viruses, approximately 200 different viruses (**TABLE 6-1**).

Although many viruses are eliminated from our bodies by the immune system, some such as those that cause fever blisters and genital herpes, take refuge in certain body cells. When we're stressed, they emerge, causing problems. The genital herpes virus, for example, produces tiny, painful sores on the genitals, thighs, and buttocks.

TABLE 6-1

How Do You Know If You Have A Cold or the Flu?

Symptoms	Cold	Flu
Onset	Gradual	Sudden
Fever	Rare	Common, may reach 101°F; may last 3 to 4 days
Headache	Rare	Common and can be severe
Cough	Hacking	Dry cough
Muscle aches and pains	Slight	Typical, often severe
Tiredness and weakness	Mild	Common, often severe
Chest discomfort	Mild to moderate	Common
Stuffy nose	Common	Sometimes
Sneezing	Usual	Sometimes
Sore throat	Common	Sometimes; may last 3 to 4 days
Caused by	Any of 200 viruses	Influenza virus

Classified as single-celled organisms, bacteria lack the organelles of more complex cells like those in humans and other animals. Capable of growing and reproducing on their own outside of cells, bacteria contain a circular strand of DNA (**FIGURE 6-2**). Outside the plasma membrane of bacteria is a thick, rigid cell wall.

Like viruses, bacteria enter the body through the respiratory tract and GI tract. They enter through the epithelium of the urinary system, as well as cuts and abrasions in the skin. Inside the body, bacteria proliferate, using body nutrients to make more of their kind. Some bacteria produce toxins—substances that cause illness and, in some cases—death.

Bacterial infections are treated with drugs called *antibiotics*. Antibiotics inhibit protein synthesis in bacteria. Although highly effective, the heavy use of antibiotics throughout the world has resulted in the formation of antibiotic-resistant strains of bacteria. Many people die each year of infections caused by these. Nationwide, one out of every four bacterial infections today involves a resistant strain. Most of these infections occur in children under the age of five. To treat them, doctors must use stronger doses or newer antibiotics.

To counteract antibiotic resistant strains of bacteria, many doctors are now being more careful about treating patients with antibiotics. They try to be sure that the infections they are treating are caused by bacteria and not by viruses, on which these drugs have not effect. For viral infections, the best remedy is usually rest, which gives the immune system time to eliminate the virus. In recent years, pharmaceutical companies have released anti-viral drugs.

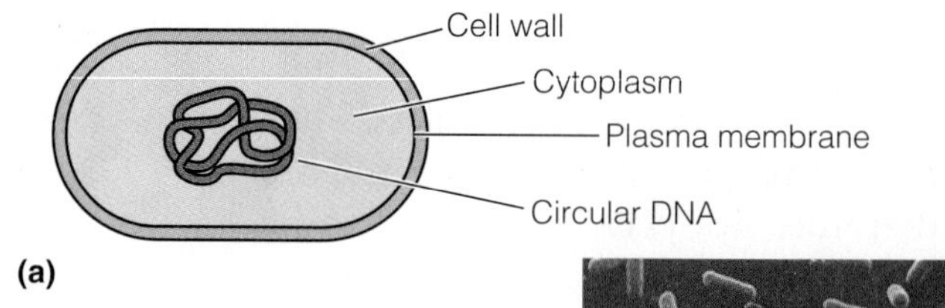

(a)

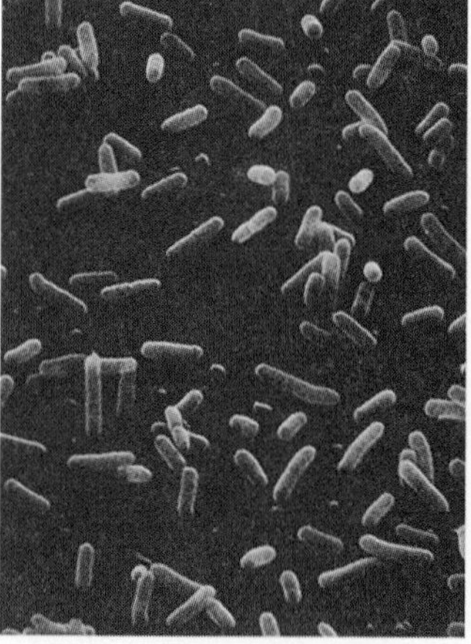

(b)

FIGURE 6-2 General Structure of a Bacterium (*a*) Bacteria come in many shapes and sizes, but all have a circular strand of DNA, cytoplasm, and a plasma membrane. Surrounding the membrane of many bacteria is a cell wall. (*b*) Electon micrograph of salmonella bacteria.

With this in mind, let's turn our attention to the ways the body protects us against harmful viruses and bacteria.

The First Line of Defense

The first line of defense against bacteria, viruses, and other microbes is the skin and the linings of the respiratory, digestive, and urinary systems. They form a physical barrier that does a decent job of preventing potentially harmful microorganisms from invading the underlying tissues. A break in these linings, however, permits microorganisms to enter.

The first line of defense also involves chemical deterrents. The skin, for instance, produces a slightly acidic secretion that impairs bacterial growth. The stomach lining releases hydrochloric acid that destroys ingested bacteria. Saliva contains an enzyme that dissolves the cell wall of bacteria, killing them. Cells in the lining of the respiratory tract produce mucus with antimicrobial properties.

The Second Line of Defense

The first line of defense is not perfect. Tiny breaks can occur, allowing viruses and bacteria to enter the body. Fortunately, there are several chemical and cellular agents that take up the battle from this point.

A cut or abrasion, for instance, results in the **inflammatory response**. Characterized by redness, swelling, pain, and heat, the inflammatory response is a kind of chemical and biological warfare waged against harmful microorganisms. It begins with the release of a variety of chemical substances by the injured tissue. Some chemicals attract cells in the tissues known as *macrophages* and certain white blood cells to the site. Here, these cells engulf or phagocytize bacteria that have entered the wound. Soon after these cells begin to work, a yellowish fluid called *pus* begins to exude from the wound. It contains dead white blood cells, microorganisms, and cellular debris.

During the inflammatory response, other chemical substances released by injured tissues stimulate repair. **Histamine** (HISS-tah-mean) is one such chemical. Released by injured tissue, histamine causes arterioles in the damaged area to expand or dilate. Dilation allows more blood to flow into the region (FIGURE 6-3). Increased blood flow is responsible for the heat and redness in the wounded area. Heat accelerates healing.

Still other substances released by injured tissues increase the leakiness of capillaries, increasing the flow of plasma into a wounded region. Oxygen and nutrients in the plasma accelerate healing.

Other cells release chemicals that cause the brain to raise body temperature, creating fever. Mild fevers cause the spleen and liver to remove iron from the blood. Because many disease-causing bacteria require iron to reproduce, fever fights the bacterial infection.

Another chemical safeguard, not part of the inflammatory response, is a group of small proteins known as the *interferons* (in-ter-FEAR-ons). **Interferons** are released from cells infected by viruses. These chemicals bind to the cell membranes of noninfected body cells. This, in turn, causes the uninfected cells to produce cellular enzymes that inhibit viral replication. If a virus enters, they cannot replicate. Interferons therefore help to halt the spread of viruses from one cell to another.

A group of blood proteins also helps fight bacterial infections. These proteins bind together to form a large structure that embeds itself in the membrane of bacteria, creating an opening into which water flows. The influx of water causes bacterial cells to swell, burst, and die.

The Third Line of Defense: The Immune System

The immune system is the third line of defense. Unlike the respiratory or digestive systems, the immune system is rather diffuse—dispersed throughout the body. Lymphocytes are a key component of the immune system. They circulate in the blood and lymph and also take up residence under the linings of the respiratory and digestive systems. Many also reside in the spleen, thymus, lymph nodes, and tonsils, known as *lymphoid organs*.

The immune response is triggered by large foreign molecules, such as proteins. These molecules are called **antigens** (AN-tah-gins), an abbreviation for antibody-generating substances. By themselves, small molecules generally do not stimulate an immune reaction. However, small molecules such as formaldehyde, penicillin, and the poison ivy toxin can bind to naturally occurring proteins in the body. The resulting combination is viewed by the body as a foreign substance.

The immune system also reacts to viruses and bacteria, and other microbes. In addition, it responds to parasites such as the organism that causes malaria. Viruses, bacteria, and parasites elicit a response because they are enclosed by membranes, or coats, that contain large molecules such as proteins or long-chained carbohydrates-that is, antigens.

Cells transplanted from one person to another

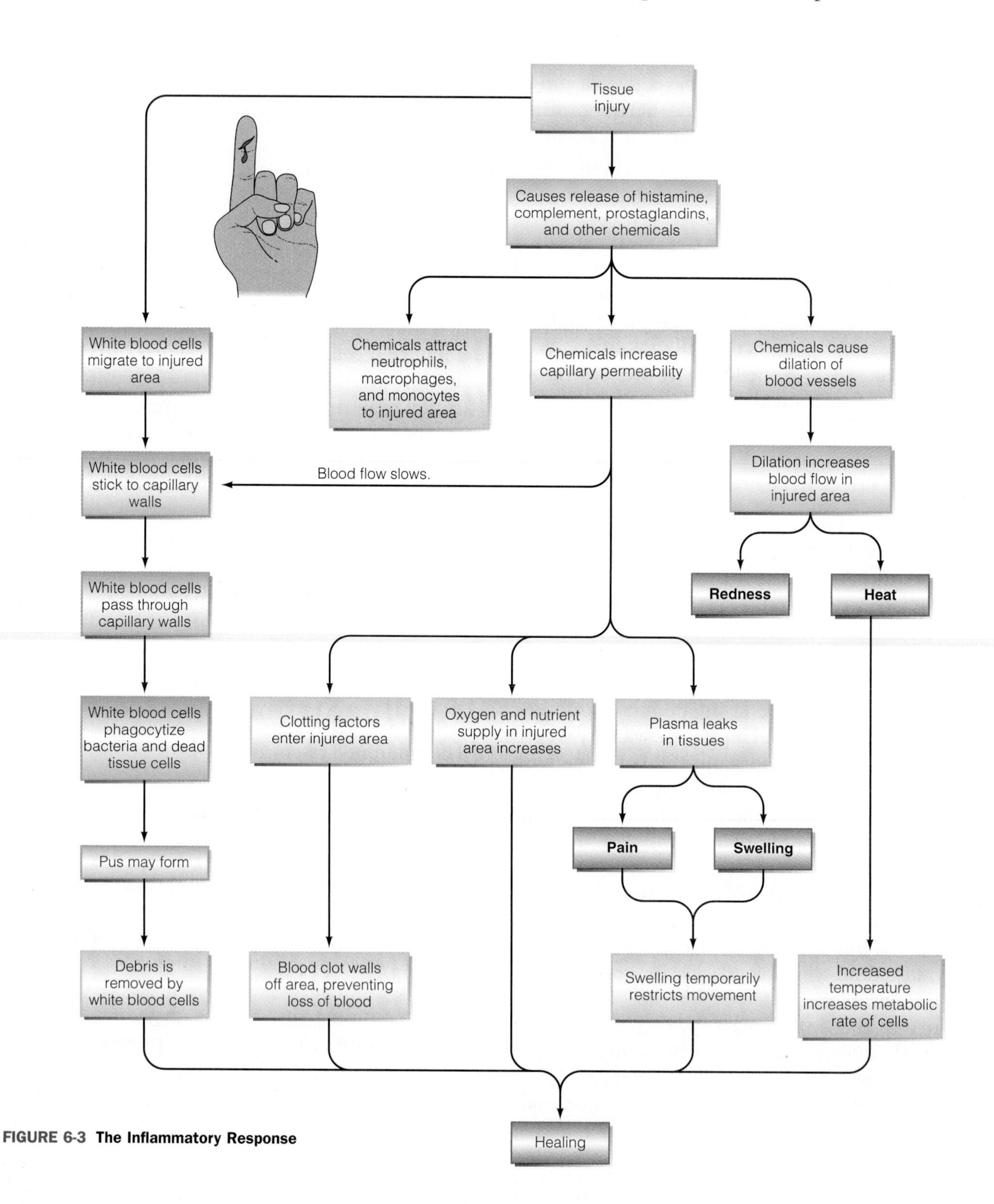

FIGURE 6-3 The Inflammatory Response

also stimulate an immune response. That's because each individual's cells contain a unique set of proteins. The immune system is activated by them.

Cancer cells also have a slightly different set of membrane proteins than normal body cells. They are therefore viewed as foreign cells in our bodies to which the immune system responds. Although cancer cells evoke an immune response, it is often not sufficient to stop the disease.

The human body contains two types of lymphocytes: **T lymphocytes**, commonly called **T cells**, and **B lymphocytes**, also called **B cells**. B cells and T cells respond to different types of antigens and function quite differently. As a rule, B cells recognize and react to bacteria, bacterial toxins, and a few viruses. When activated, B cells produce antibodies to these antigens.

T cells recognize and respond to our own body cells that have gone awry. This includes cancer cells as well as body cells that have been invaded by viruses. T cells also respond to transplanted tissue cells and larger disease-causing agents, such as single-celled fungi and parasites. Unlike B cells, T cells attack their targets directly.

By various estimates, several million distinct B and T cells are produced in the body early in life. Each one is programmed to respond to a specific antigen. Over a lifetime, only a relatively small fraction of these cells will be called into duty.

How B Cells Work

When a bacterium first enters the body, B cells programmed to respond to the unique proteins found in the bacterium's cell membrane bind to it. The B cells begin to divide. Some of the cells produced in this process form a new kind of cell, called the **plasma cell**. Plasma cells produce antibodies, proteins that help eliminate antigens, whether they're entire bacteria or bacterial toxins.

Antibodies circulate in the blood and lymph, where they bind to antigens that triggered the response. As noted above, bacteria, bacterial toxins, and some viruses are their main targets of B cells.

The first time an antigen enters the body, it elicits an immune response, but the initial reaction—or **primary response**—is relatively slow (FIGURE 6-4A). During the primary response, antibody levels in the blood do not begin to rise until approximately the beginning of the second week after the intruder was detected, partly explaining why it takes most people about 7-10 days to combat a cold or the flu. This delay occurs because it takes time for B cells to form a sufficient number of plasma cells.

If the same antigen enters the body at a later date, however, the immune system acts much more quickly (FIGURE 6-4B). This stronger reaction constitutes the **secondary response**. During a secondary response, antibody levels increase rather quickly—a few days after the antigen has entered the body. The amount of antibody produced also greatly exceeds quantities produced during the primary response. Consequently, the antigen is quickly destroyed, and a recurrence of the illness is prevented.

Why is the secondary response so much faster? The reason is that during the primary response, some lymphocytes divide to produce a large population of **memory cells**. Memory cells remain

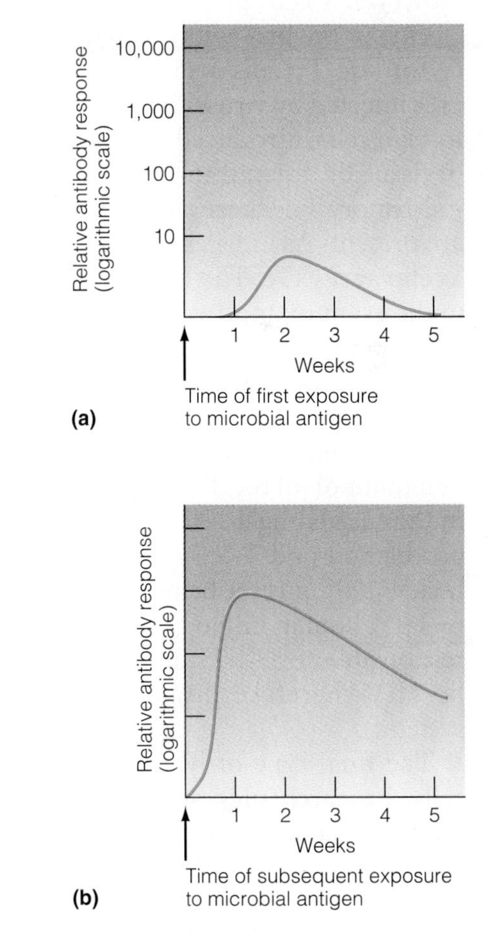

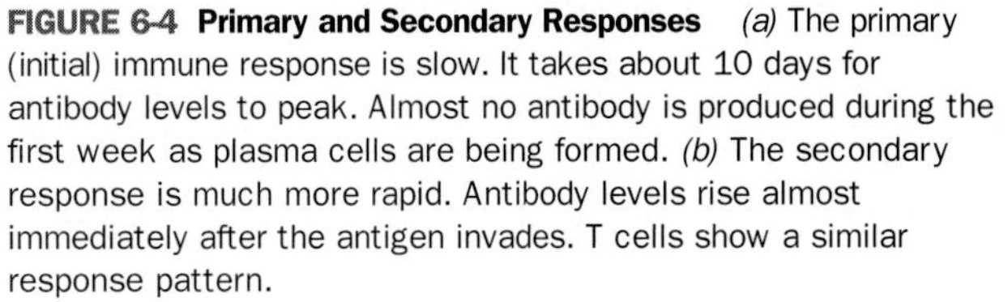
FIGURE 6-4 Primary and Secondary Responses *(a)* The primary (initial) immune response is slow. It takes about 10 days for antibody levels to peak. Almost no antibody is produced during the first week as plasma cells are being formed. *(b)* The secondary response is much more rapid. Antibody levels rise almost immediately after the antigen invades. T cells show a similar response pattern.

in the body awaiting the antigen's reentry. When it reenters, they divide rapidly, producing numerous plasma cells that quickly crank out antibodies to combat the foreign invaders.

Antibodies destroy foreign organisms and antigens via one of several ways. Some antibodies, for instance, bind to antigens, forming a complete coating around them. This prevents the antigen from doing any harm. Other antibodies bind to antigens and cause them to clump together, again rendering them ineffective. Still others bind to antigens, forming much larger, water-insoluble complexes. These precipitate or fall out of solution.

How T Cells Work

Like B cells, T cells respond to the presence of antigens by undergoing rapid cell division. However, T cells produce four different cells.

One of the products is called *Cytotoxic T cells*. **Cytotoxic T Cells** attack and kill viruses, body cells infected by viruses, parasites, cancer cells, and foreign cells introduced during blood transfusions or tissue or organ transplants.

Memory T cells are also produced when antigens are present. As in B cells, the memory T cells form a cellular reserve. They're there to protect the body in the event of another invasion.

Helper T cells are also produced when the body detects an antigen. Helper T cells greatly enhance the immune response by stimulating both B cells and cytotoxic T cells. Helper T cells are the most abundant of all the T cells (comprising 60%-70% of the circulating T cells). Without them, antibody production and T-cell activity is greatly reduced. Incidentally, it is the helper T cells that are targeted by HIV, human immunodeficiency virus, which is responsible for AIDS. Infection of helper T cells by HIV therefore disables a person's immune system.

The final type of lymphocyte is **suppressor T cells**. Research suggests that they "turn off" the immune reaction as the antigen begins to disappear.

Active and Passive Immunity

One of the major medical advances of the 1800s was the discovery of **vaccines** (vac-SEENS). Used to prevent bacterial and viral infections, vaccines contain inactivated or greatly weakened viruses, bacteria, or bacterial toxins. When injected into the body, the "disabled" antigens in vaccines elicit an immune response just as if the real thing had entered the body. The immune system responds to the virus by producing antibodies or T cells.

Vaccines stimulate the immune reaction because the weakened or deactivated organisms (or toxins) they contain still possess the antigenic proteins or carbohydrates that trigger B- and T-cell activation. However, because the infectious agents have been seriously weakened or deactivated, viruses, bacteria, and bacterial toxins in vaccines do not cause disease. Some individuals may develop minor symptoms, but they're not life threatening.

Vaccination provides a form of protection that immunologists call **active immunity**—so named because the body actively produces memory T and B cells to protect a person against future infections. The immune response is the same as the one that occurs when a disease-causing organism enters the body. Most vaccines provide immunity or protection from microorganisms for long periods, sometimes for life. Others need to be administered several times during one's lifetime.

Vaccinations are vital in controlling deadly diseases such as polio, typhus, and smallpox—diseases that can cripple or kill people before their immune system mounts an effective response.

The second type of immunity, called **passive immunity**, is a temporary form of protection. It results from the injection of antibodies. These antibodies are produced by injecting antigens in other animals such as sheep. The antibodies are then extracted from the blood for use in humans. Passive immunity is so named because the cells of the immune system are not activated. T cells and B cells are not called into duty.

Antibodies remain in the blood for a few weeks, protecting an individual from infection. Because the liver slowly removes these molecules from the blood, a person gradually loses protection.

Antibodies are administered to prevent illnesses. Travelers to developing nations, for instance, are often given antibodies to viral hepatitis (liver infection) as a preventive measure. Antibodies are also used to treat individuals who have been bitten by poisonous snakes to counteract the venom (**FIGURE 6-5**). Venom is a mixture of proteins, enzymes, and polypeptides (long-chain molecules made up of amino acids, but not long enough to be classified as a protein). These molecules damage body cells, especially nerve cells and heart muscle cells. Bites of poisonous snakes can be treated by antivenom. Antibodies destroy or deactivate the harmful molecules in snake venom before they can have adverse effects.

Vaccines have lowered the incidence of many infectious diseases in the United States and other industrialized nations by 99% or more. Despite the success of vaccines, publicity concerning their rare side effects has caused many parents to choose not to have their children vaccinated. In addition, because immunization programs have greatly reduced the incidence of most infectious diseases, some parents feel their children are safe without vaccines.

Public health officials are quick to point out that disease-causing microorganisms that once took a huge toll on humans have not been eradicated. Without continued protection, outbreaks could occur again.

Blood Transfusions and Tissue Transplantation

Although the immune system is important in protecting us from microorganisms, it can work against doctors when they perform life-saving blood transfusions and tissue transplants.

Blood transfusions require careful cross-matching of donors and recipients to be certain their blood types match. Humans have four blood types: A, B, AB, and O. The letters refer to the type of antigens found on the surface of the red blood cells. In type A blood, there's A antigen. In Type B, there's B antigen. In Type AB both antigens are present. In Type O there are no antigens. Cross-matching blood is essential to prevent life-threatening immune reactions.

Organ and tissue transplantation is a much more complex matter. Only three conditions exist in which a person can receive a transplant and not reject it. One is if the tissue comes from an individual's own body. For burn victims, surgeons might use healthy skin from one part of the body to cover a badly damaged region elsewhere. Hair transplants also qualify when hair from the back of the head is moved to balding spots.

FIGURE 6-5
Poison and Antidote Poisonous snakes like this rattler inject venom into their victims. Venom can be milked from the snake and used to produce antivenom, a serum containing antibodies that neutralize the venom.

The second instance is when a tissue is transplanted between identical twins—individuals derived from a single fertilized egg that splits early in embryonic development to form two embryos. These individuals are genetically identical and have identical cell membrane proteins.

A third instance occurs when tissue rejection is inhibited by drugs. For example, in heart and kidney transplants treatment with these drugs must be continued throughout the life of the patient. Unfortunately, most of these drugs leave patients vulnerable to bacterial and viral infections.

Diseases of the Immune System

The immune system, like all other body systems, can malfunction. One of the most common malfunctions results in allergies.

An **allergy** is an overreaction to some antigens such as pollen or a food (**FIGURE 6-6**). Antigens that stimulate allergic reactions are called **allergens** (AL-er-gens). Allergens cause the production of one class of antibodies, the **IgE** antibodies, from plasma cells. As Figure 6-6 shows, these antibodies bind to a type of cell known as a *mast cell*. Found in many tissues, mast cells contain large quantities of histamine.

When allergens bind to the IgE antibodies attached to the mast cells, mast cells release massive amounts of histamine (Figure 6-6). Histamine released in the lungs causes tiny tubules carrying air

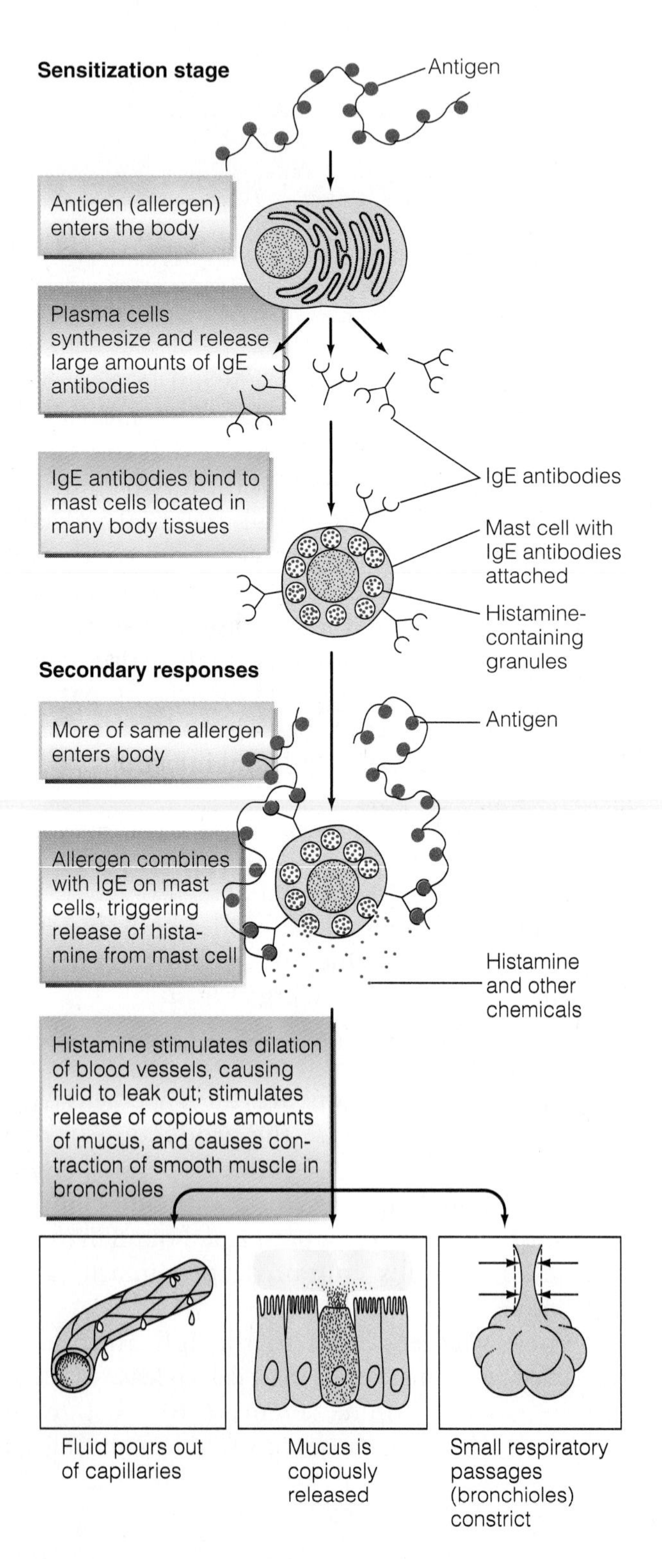

FIGURE 6-6 Allergic Reaction Antigen stimulates the production of massive amounts of IgE, a type of antibody produced by plasma cells. IgE attaches to mast cells. This is the sensitization stage. When the antigen enters again, it binds to the IgE antibodies on the mast cells, triggering a massive release of histamine and other chemicals. Histamine, in turn, causes blood vessels to dilate and become leaky. This triggers the production of mucus in the respiratory tract. In some people, the chemicals released by the mast cells also cause the small air-carrying ducts in the lungs to constrict, making breathing difficult.

in the lung to constrict. This reduces airflow and makes breathing more difficult. This condition is called **asthma** (AS-ma).

Allergic reactions usually occur in specific body tissues, where they create local symptoms that, while irritating, are not life-threatening. For example, an allergic response may occur in the eyes, causing redness and itching. Or, it may occur in the nasal passageway, causing stuffiness.

An allergic response can occur in the bloodstream, where it can be fatal if not treated quickly. For example, penicillin or bee venom in the bloodstream of certain people can cause a massive release of histamine and other chemicals. This, in turn, causes extensive dilation of blood vessels in the skin and other tissues. The blood pressure then falls, shutting down the circulatory system. Histamine released by mast cells also causes severe constriction of the ducts in the lungs that deliver oxygen to the air sacs, making breathing difficult. The decline in blood pressure and constriction of the bronchioles result in **anaphylactic shock** (ANN-ah-fah-LACK-tic). Death may follow if measures are not taken quickly. One such measure is an injection of the hormone epinephrine (commonly known as adrenalin), which rapidly reverses the constriction of the bronchioles.

Allergies can be treated by avoiding allergens—for instance, avoiding milk or staying clear of dogs and cats. Patients can also take antihistamines (an-tea-HISS-tah-means), drugs that counteract the effects of histamine. Patients may also be given allergy shots, injections of increasing quantities of the allergen to which they're allergic. In many cases, this treatment makes an individual less and less sensitive to the allergen.

Occasionally, the immune system really runs amok, mounting an attack on the body's own cells. This unfortunate state of affairs is known as an *autoimmune disease*. **Autoimmune diseases** result from many causes. For example, in some instances, normal body proteins can be modified by environmental pollutants, viruses, or genetic mutations so that they are no longer recognizable by the body.

AIDS: The Deadly Virus

Many millions of people the world over have died from a disease known as *AIDS* (acquired immune deficiency syndrome). AIDS is caused by a virus that attacks and weakens the immune system. This virus, known as **human immunodeficiency virus** or **HIV** for short, attacks the helper T cells, severely impairing the immune system. Patients die from infections.

HIV is transmitted by sexual contact, usually from infected men to uninfected men and women. It is also transmitted in blood transfusions and by sharing hypodermic needles among drug addicts. It can also be transmitted from infected mothers to their babies at birth.

The number of cases of AIDS and the number of deaths from AIDS in the United States increased dramatically since the early 1980s. HIV is a global epidemic with more than 42 million people infected worldwide in 2002 and approximately 5 million new infections each year. Most, if not all, of these people will die from the disease, which is still spreading rapidly through different countries such as Africa and China and different populations, such as African-Americans, in the United States.

AIDS progresses through three stages. During the first phase, no symptoms appear, although an individual is highly infectious—able to transmit the disease to others. During the second phase, patients grow progressively weaker as their immune systems falter. Lymph nodes swell and patients report persistent or recurrent fevers and persistent coughs. Mental deterioration may also occur. During the last phase, patients suffer from severe weight loss and weakness. Many develop cancer and bacterial infections because of their diminished immune response (**FIGURE 6-7**).

Stopping the virus has proved difficult, in large part because symptoms of AIDS do not appear until several months to several years after the initial HIV infection. Abstinence and the use of condoms are both effective in helping to stop the spread of the disease. Several drugs have also been developed that slow down the progression of the disease. Used in combination, they are greatly prolonging the lives of those infected with HIV, although treatment is extremely expensive. Numerous researchers are developing vaccines which they hope will protect people and eventually eradicate the virus.

The immune system is one of our greatest allies in maintaining health. It is called into duty every day to protect us from danger. But like other body systems, it can run amok. It is not invincible. You can, however, increase your chances of maintaining a strong immune system by eating right, getting plenty of sleep, minimizing stress, and living a healthy lifestyle.

FIGURE 6-7 HIV and Kaposi's Sarcoma *(a)* AIDS viruses. *(b)* Kaposi's sarcoma on foot found only in AIDS patients.

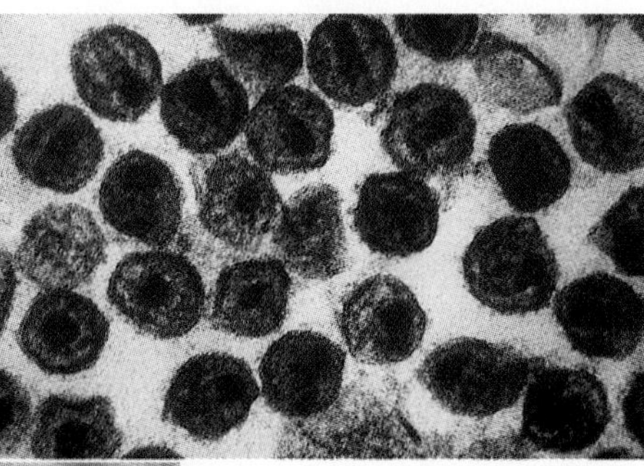

(a)

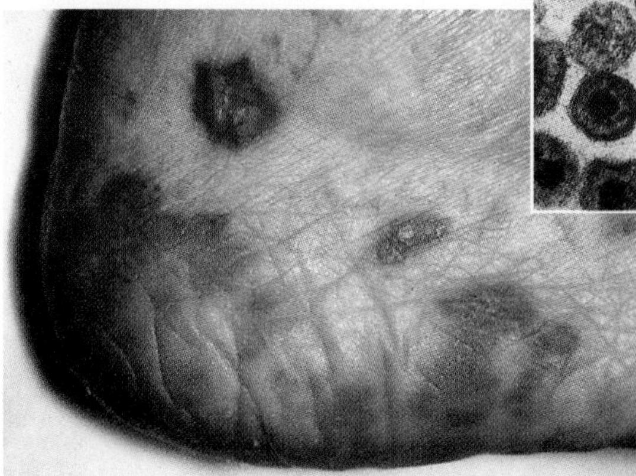

(b)

The Respiratory System

The human respiratory system functions automatically, drawing air into the lungs, then letting it out. This cycle repeats itself about 16 times per minute at rest—or about 23,000 times per day. Like so many other systems, the respiratory system is vital to our survival. Not only does it supply the blood with oxygen required by cells to make energy, it helps get rid of carbon dioxide, a waste product of cellular energy production.

In this module, we will explore the structure and function of the respiratory system—and some of the diseases that affect it.

Structure and Function of the Respiratory System

The respiratory system consists of two parts: one part that transports air into and out of the lungs and a second part in which gas-exchange occurs between the air we breathe and our blood stream.

The air-conducting portion of the respiratory system is an elaborate set of passageways that transports air to and from the lungs, two large, saclike organs in the thoracic cavity (**FIGURE 7-1A**). Like the arteries of the body, these passageways start out large then become progressively smaller and more numerous, branching extensively in the lungs.

The lungs are the gas-exchange portion of the respiratory system. Each lung contains millions of tiny, thin-walled air sacs called **alveoli** (al-vee-oh-lie) (**FIGURE 7-1B**). The walls of the alveoli are surrounded by numerous capillaries that absorb oxygen from the inhaled air and release carbon dioxide (Figure 7-1B).

Air enters the respiratory system through the nose and mouth, then is drawn backward through the nasal cavity into a structure known as the *voice box* or *larynx* (LAIR-inks). The **larynx** is a rigid, hollow structure that houses the **vocal cords**, two folds of tissue that vibrate when we talk, sing, or hum (**FIGURE 7-2**). From here, air empties into the windpipe or **trachea** (TRAY-kee-ah).

As explained in Module 3, food is prevented from entering the larynx by the **epiglottis** (ep-eh-GLOT-tis), a flap of tissue that closes off the opening to the larynx during swallowing. Occasionally, however, food accidentally enters the larynx. This leads to

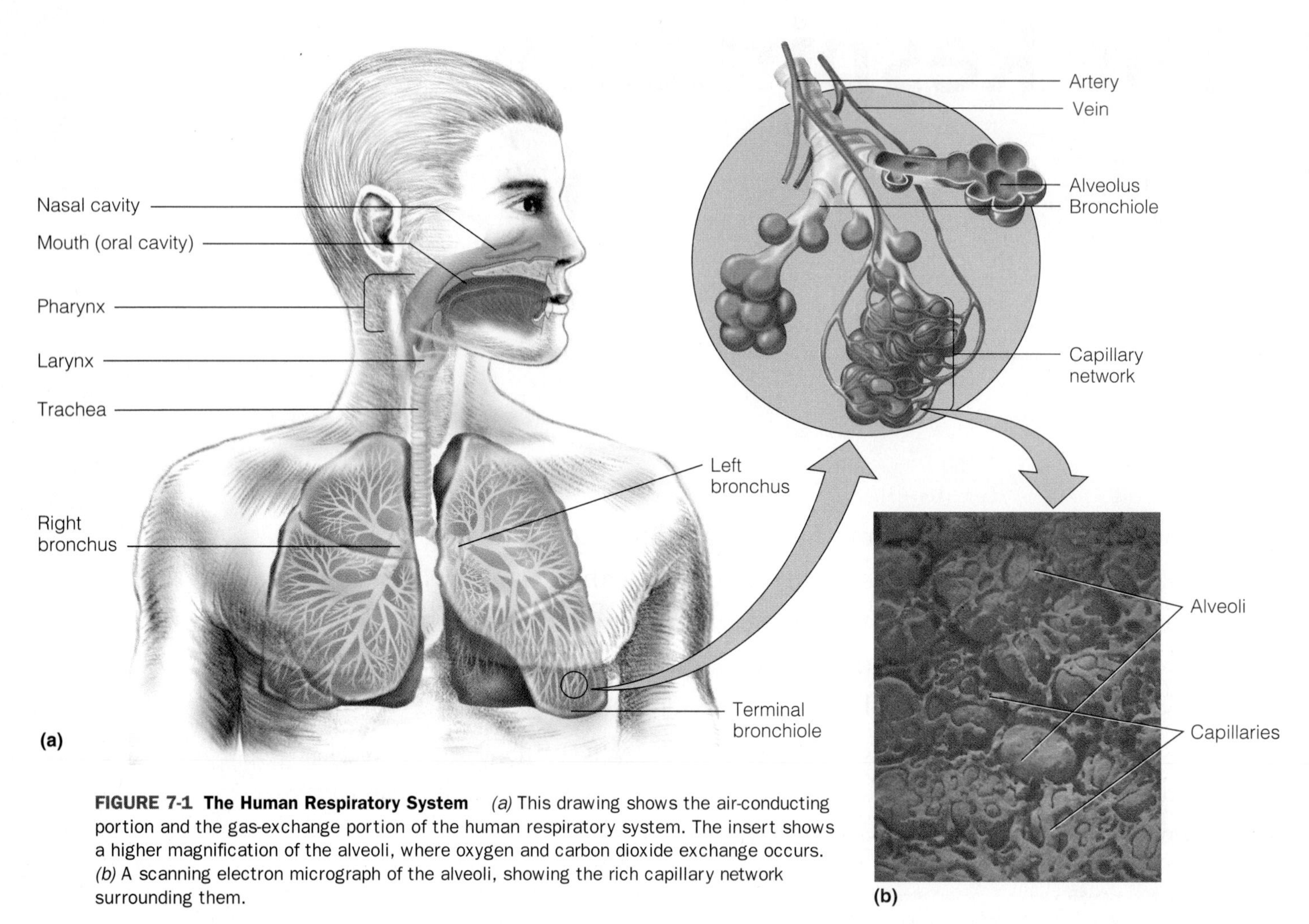

FIGURE 7-1 The Human Respiratory System *(a)* This drawing shows the air-conducting portion and the gas-exchange portion of the human respiratory system. The insert shows a higher magnification of the alveoli, where oxygen and carbon dioxide exchange occurs. *(b)* A scanning electron micrograph of the alveoli, showing the rich capillary network surrounding them.

violent coughing. Coughing is a reflex that helps eject the food and keep us from suffocating.

The trachea enters the chest cavity, where it divides into two large branches. They enter the lungs alongside the arteries and veins. Inside the lungs, these tubes branch extensively, forming progressively smaller tubules that carry air to the alveoli.

The smallest of these tubules branch to form bronchioles (BRON-kee-ols). **Bronchioles** are small ducts that lead directly to the alveoli. Smooth muscle in the walls of the bronchioles permits the bronchioles to open and close, and thus provides a means of controlling air flow in the lungs. During exercise or times of stress, the bronchioles open and the flow of air into the lungs increases. This homeostatic mechanism helps meet the body's need for more oxygen.

The conducting portion of the respiratory system removes airborne particles, such as dust, in the air we breathe. The larger and medium sized particles drop out as inhaled air travels through the nasal cavity and passageways of the upper respiratory system. Smaller particles, also known as *fine particulates*, however, are so small that they remain in the air breathed into our lungs. As a result, they can penetrate deeply into the lungs. Some of these particles contain toxic metals such as mercury, which can cause lung cancer.

Particles that precipitate out of the inhaled air in the upper portion of the respiratory system are trapped in a layer of mucus (MEW-kuss). Mucus is a thick, slimy secretion produced by certain cells in the epithelial lining of the upper respiratory tract (**FIGURE 7-3**).

The epithelium of the respiratory tract also contains numerous ciliated cells. **Cilia** are tiny organelles that extend from the surface of cells. They beat upward, transporting mucus containing bacteria and dust particles to the mouth. When the

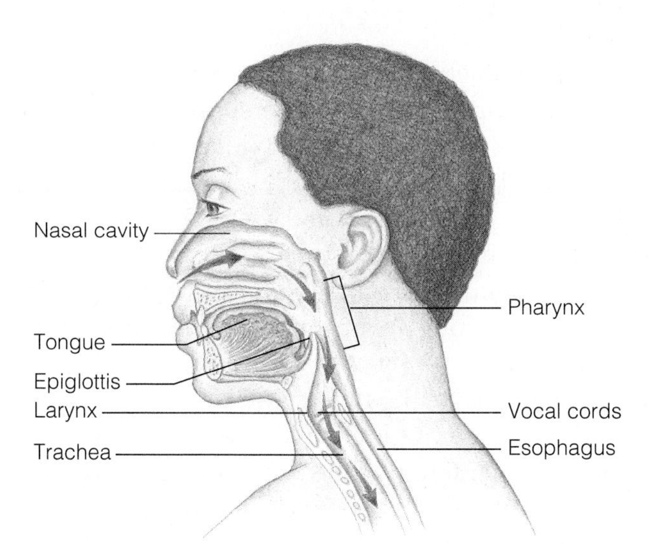

FIGURE 7-2 Uppermost Portion of the Respiratory System Bony protrusions into the nasal cavity (not shown here) create turbulence that causes dust particles to settle out on the mucous coating. Notice that air passing from the pharynx enters the larynx. Food is kept from entering the respiratory system by the epiglottis, which covers the laryngeal opening during swallowing.

mucus reaches the mouth, it may be swallowed or spit out. This mechanism protects the respiratory tract and lungs from bacteria and potentially harmful particulates.

Like all homeostatic mechanisms, the respiratory mucous trap is not invincible. Bacteria and viruses do occasionally penetrate the lining, causing respiratory infections. In addition, sulfur dioxide in cigarette smoke temporarily paralyzes, and may even destroy, cilia. Sulfur dioxide gas in the smoke of a single cigarette paralyzes the cilia for an hour or more, permitting bacteria and toxic particulates to be deposited in the respiratory tract, even enter the lungs. Ironically, the cilia of a smoker are paralyzed when they are needed the most!

Because of this, smokers suffer more frequent respiratory infections than nonsmokers. Research shows that alcohol also paralyzes the respiratory system cilia, explaining why alcoholics are also prone to respiratory infections.

Beneath the epithelium of the respiratory tract is a rich network of capillaries that releases heat and moisture. As air passes through the conducting portion it is therefore warmed and moistened. Moisture protects the lungs from drying out, and heat protects them from cold temperatures. By the time inhaled air reaches the lungs, it is nearly saturated with water and is warmed to body temperature, except in extremely cold areas.

The Alveoli. Air consists principally of nitrogen and oxygen. It also contains small amounts of carbon dioxide and other gases. Oxygen in the atmosphere is generated by photosynthesis in plants and algae. A constant supply must be delivered to body cells to sustain cellular energy production.

Oxygen is transported to the lungs in the air we breathe. Inside the estimated 150 million alveoli of our lungs, oxygen escapes into the blood stream. The alveoli are lined by a single layer of flattened cells and are surrounded by an extensive network of capillaries, shown in Figure 7-1B. Each capillary is made of an equally thin layer of flattened cells. Together, the cells of the alveolar wall and the capillary present a fairly easy route for the passage of gases into and out of the alveoli.

Oxygen travels from the alveoli to the blood stream by diffusion. It then diffuses across the cell membrane of red blood cells where it attaches to

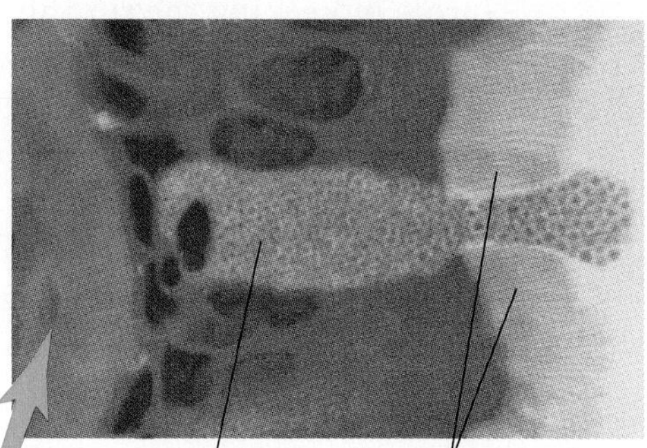

FIGURE 7-3 Mucous Trap *(a)* Drawing of the lining of the trachea. Mucus produced by the mucous cells of the lining of much of the respiratory system traps bacteria, viruses, and other particulates in the air. The cilia transport the mucus toward the mouth. *(b)* Higher magnification of the lining showing a mucous cell and ciliated epithelial cells.

hemoglobin molecules. It is then transported via the arteries to the cells, tissues, and organs of the body.

Carbon dioxide travels in the opposite direction, also by diffusion, leaving the blood and entering the alveoli. When we exhale, carbon dioxide is released into the air.

Inside the lungs in the alveoli are cells called **alveolar macrophage**, also known as the **dust cells**. Alveolar macrophages wander freely through the alveoli, gobbling dust, bacteria, viruses, and other particulates that escaped filtration in the upper portions of the respiratory system (**FIGURE 7-4**). Once filled, they retire to the sidelines, taking up residence in the connective tissue surrounding the alveoli. Because there are so many particulates in tobacco smoke, a smoker's lungs are often blackened by dust cells packed with particulates. The lungs of urban residents may also be blackened by the accumulation of smoke and dust particles.

The walls of the alveoli also contain large, round cells that produce a chemical known as **surfactant** (sir-FACK-tant). It dissolves in the thin layer of water lining the alveoli where it plays an important role in keeping the tiny alveoli from collapsing.

Some premature babies lack sufficient surfactant. As a result, the larger alveoli collapse, making it difficult to breath. The condition, known as *respiratory distress syndrome*, is usually treated with an artificial surfactant. It keeps the alveoli open until the lungs produce enough of their own.

Making Sounds

The chief functions of the respiratory system are to replenish the blood's oxygen supply and rid the blood of excess carbon dioxide. However, the respiratory system serves other functions as well. The vocal cords, located in the larynx, for example, produce sounds that allow people to communicate.

The sounds we make from grunts and groans to speech to song are produced by vibration of the vocal cords. The **vocal cords** are two elastic ligaments inside the larynx (**FIGURE 7-5**). The vocal cords vibrate as air is expelled from the lungs. The sounds generated by the vocal cords are modified by changing the position of the tongue and the shape of the oral cavity.

The vocal cords vary in length and thickness from one person to the next, which accounts for differences in our voices. They also vary between men and women. Most men, for example, have longer, thicker vocal cords than women as a result of testosterone, the male sex hormone produced by the testes. Men therefore tend to have deeper voices than most women.

Bacterial and viral infections of the larynx can cause the vocal cords to swell. This thickens the cords, causing a person's voice to lower. This condition is known as **laryngitis** (lair-in-JITE-iss). Laryngitis may also be caused by tobacco smoke,

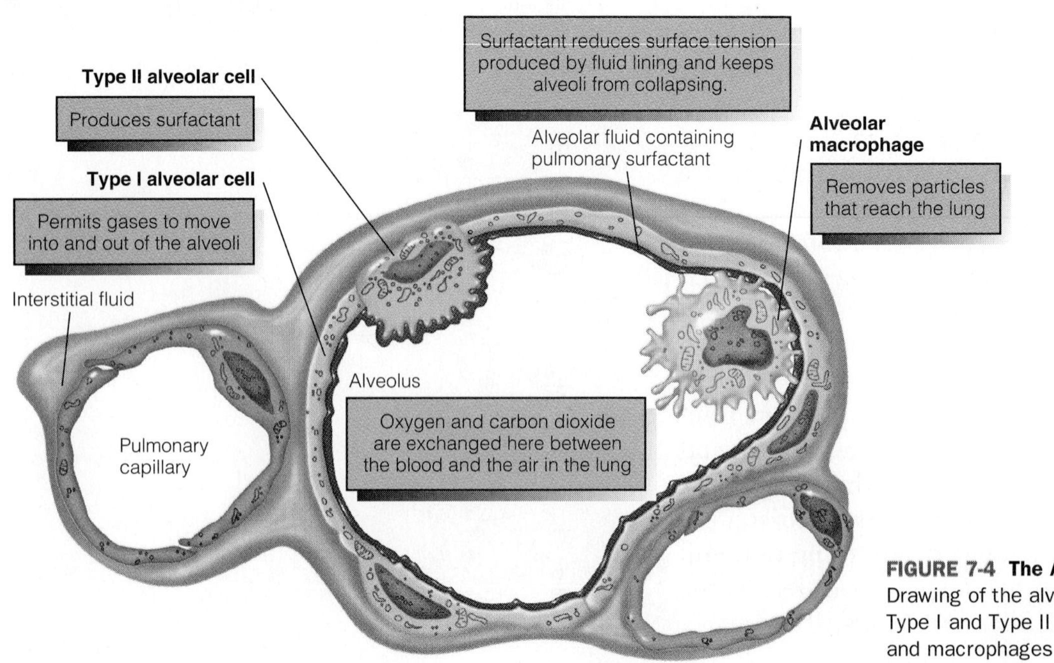

FIGURE 7-4 The Alveolar Macrophage Drawing of the alveolus showing Type I and Type II alveolar cells and macrophages or dust cells.

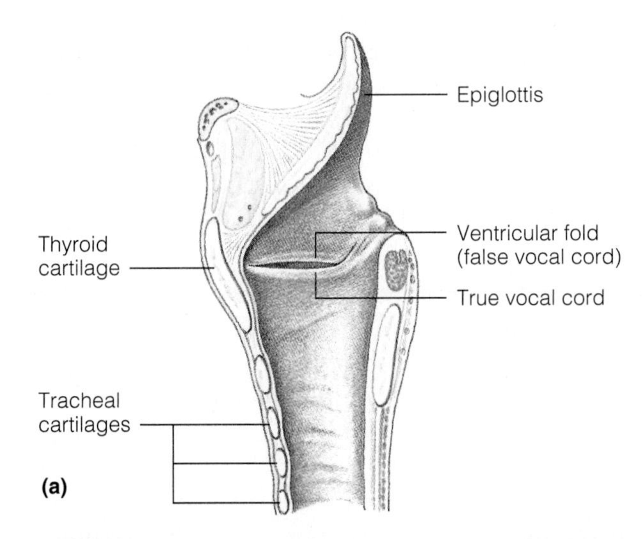

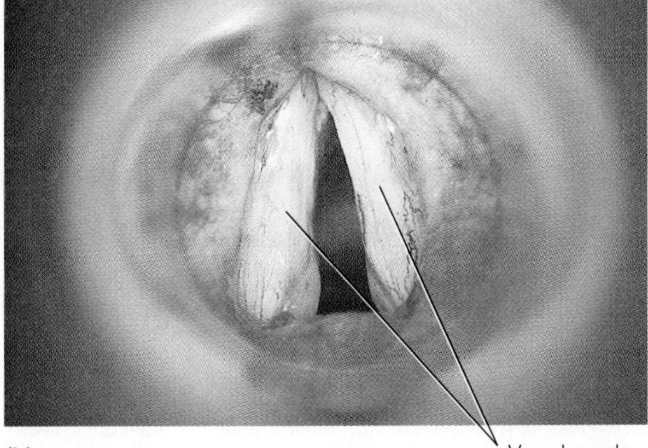

FIGURE 7-5 Vocal Cords *(a)* This drawing of the larynx shows the location of the vocal cords. Note the presence of the false vocal cord, so named because it does not function in sound production. *(b)* View into the larynx of a patient showing the true vocal cords from above.

alcohol, excessive talking, shouting, coughing, or singing—activities that can irritate the vocal cords.

The respiratory system also houses the receptors for smell. They are located in the epithelium in the roof of the nasal cavity. You'll learn more about this in Module 10.

Breathing and the Control of Respiration

Air moves in and out of the lungs in much the same way that it moves in and out of the bellows that blacksmiths use to fan their fires. Breathing, however, is largely an involuntary action, controlled by the nervous system.

To begin, air must first be drawn into the lungs. This process is known as **inhalation**. Inhalation is followed by **exhalation**, expulsion of the air.

Inhalation is controlled by the brain. Nerve impulses traveling from the brain stimulate the **diaphragm**, a dome-shaped muscle that separates the abdominal and chest cavities (FIGURE 7-6A). These impulses cause the diaphragm to contract. When it contracts, it flattens and lowers. This, in turn, draws air into the lungs in much the same way that pulling the plunger of a syringe out draws air into the device.

Inhalation also involves the short, powerful muscles that lie between the ribs. Nerve impulses traveling to these muscles cause them to contract as the diaphragm is lowered. This contraction lifts the rib cage up and out.

Together, the contractions of the diaphragm and the muscles between the ribs increases the volume of the chest cavity (FIGURE 7-6B). Air naturally flows in through the mouth and nose. The lungs expand

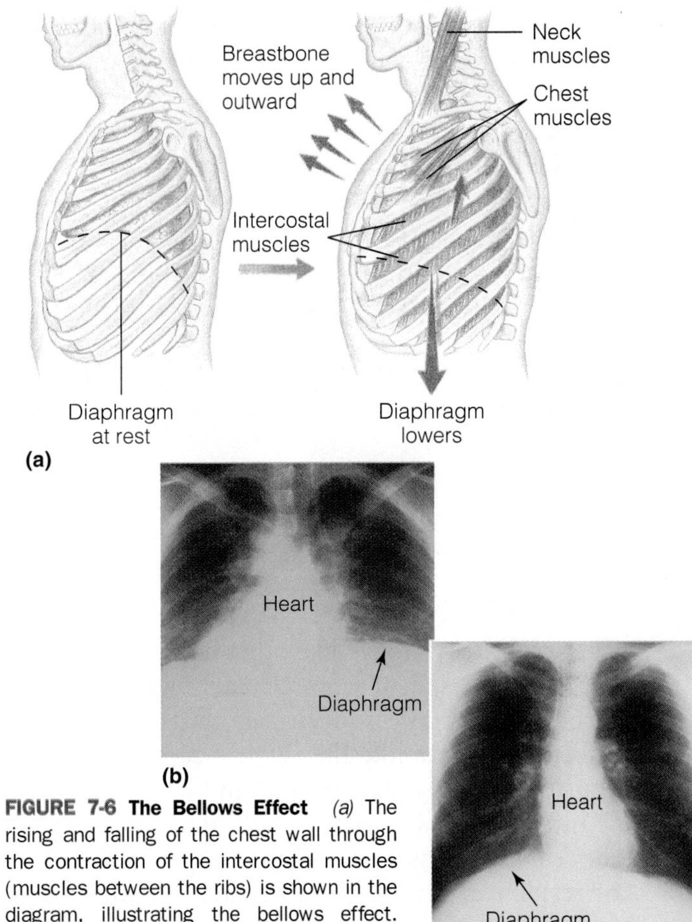

FIGURE 7-6 The Bellows Effect *(a)* The rising and falling of the chest wall through the contraction of the intercostal muscles (muscles between the ribs) is shown in the diagram, illustrating the bellows effect. Inspiration is assisted by the contraction of the diaphragm. The rising of the chest wall and the lowering of the diaphragm draws air into the lungs. *(b)* X-rays showing the size of the lungs in full exhalation *(top)* and full inspiration *(bottom)*.

like balloons. At rest, each breath delivers about 500 milliliters of air to the lungs.

Exhalation begins after the lungs have filled with air. When this occurs, the diaphragm and rib muscles relax, returning to their previous state. Elastic fibers in the walls of the lung cause the lungs to recoil. These changes reduce the volume of the chest cavity, forcing the air out.

Although exhalation does not involve contraction of muscles in an individual at rest, muscles can be called into action to force air out more quickly and more completely. Inhalation can also be consciously increased by a forceful contraction of the diaphragm and the muscles of the ribs. This increases the amount of air entering your lungs. Athletes often actively inhale and exhale just before an event to increase oxygen levels in their blood.

Breathing is controlled by a region of the brain called the *breathing center*. It contains nerve cells that produce periodic impulses. These impulses stimulate inhalation by causing the periodic contraction of the rib muscles and the diaphragm. When the lungs fill, however, the nerve impulses cease and the muscles relax. Air is forced out of the lungs.

Breathing is controlled by other processes, too. Chemical receptors inside the brain monitor carbon dioxide concentration. When levels of carbon dioxide are high, for instance, during vigorous exercise, the receptors send impulses to the breathing center. This causes an increase in the depth and rate of breathing. Next time you run up a flight of stairs and start breathing hard, you'll know why.

Diseases of the Respiratory System

The respiratory system like other body systems can become diseased. It is particularly vulnerable to airborne bacteria and viruses that can lead to infections. Some like the flu and colds can cause considerable discomfort and can be fatal.

Infections may occur in different locations in the respiratory system and are named by their site of residence. An infection in the large tubes that branch from the trachea, the bronchi, is known as **bronchitis** (bron-KITE-iss). An infection of the sinuses is known as **sinusitis** (sigh-nu-SITE-iss). Once inside the respiratory tract, bacteria, viruses, and other microorganisms can spread to other organ systems such as the brain.

The lungs are also susceptible to pollutants, including airborne materials, such as asbestos fibers. Asbestos is a naturally occurring fiber that has been used in thousands of products from pipe and sound insulation to car brake pads. Inside the lung, asbestos can cause enormous damage. Some workers exposed to asbestos, for example, develop a fatal lung cancer. Others have developed a debilitating disease known as *asbestosis* (ass-bes-TOE-sis). Asbestosis is a build-up of scar tissue that reduces lung capacity. Because asbestos is so dangerous, virtually all of its uses have been banned in the United States, and asbestos used for insulation and decoration is being removed from buildings or stabilized so it won't flake off.

Another common disease of the respiratory system is asthma. **Asthma** is a type of allergy. Characterized by periodic episodes of wheezing and difficult breathing, asthma persists for many years. Most cases of asthma are caused by allergic reactions to common allergens such as dust, pollen, and skin cells (dander) from pets. In some individuals, asthma is brought on by certain foods, such as eggs, milk, chocolate, and food preservatives. Still other cases are triggered by drugs, such as antibiotics.

In asthmatics, irritants can cause a rapid increase in the production of mucus by the bronchi and bronchioles. Irritants also stimulate the constriction of the bronchioles. Mucus production and constriction of the bronchioles make breathing difficult. Asthmatics also suffer from a chronic inflammation of the lining of the respiratory tract.

Although asthma is fairly common in school children, it often disappears as they grow older. However, studies show that asthma is starting to appear in adults. In addition, many more children suffer from the disease, too.

Public health officials are concerned about the growing epidemic of asthma. Attacks can be quite disabling. Some even lead to death. Victims are generally elderly individuals who are suffering from other diseases.

The severity of asthma attacks can be greatly lessened by proper medical treatment. One of the most common treatments is an oral spray (inhalant) containing the hormone adrenalin (epinephrine), which stimulates the bronchioles to open. Anti-inflammatory drugs (steroids) can be administered to treat chronic inflammation. Screening tests can help a patient find out what substances trigger an asthmatic attack so they can be avoided.

Another common disease is lung cancer. Each year, lung cancer claims the lives of an estimated 144,000 men and women in the United States (**FIGURE 7-7**). Lung cancer is primarily due to smoking. In fact, smokers are 11-25 times more likely to develop lung cancer than nonsmokers. The more one smokes, the more risk one suffers.

The respiratory system is vital to human life, but like other body systems it must be well taken care of to ensure a good, healthy life. Living and working in an unpolluted environment and avoiding tobacco smoke are keys to keeping the lungs and the rest of the respiratory system in peak condition.

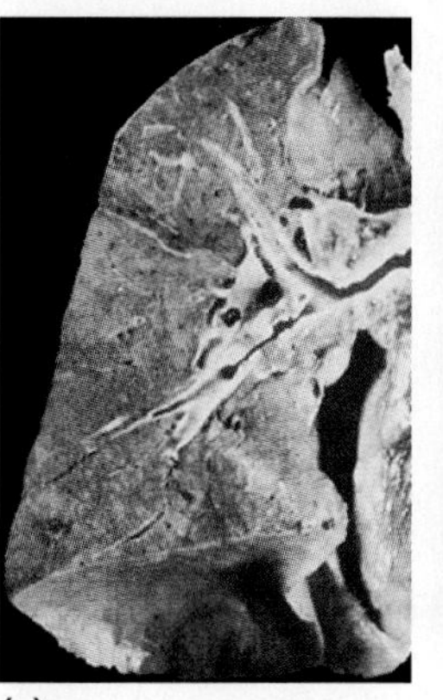

FIGURE 7-7 The Normal and Cancerous Lung *(a)* The normal lung appears spongy. *(b)* The cancerous lung from a smoker is filled with particulates and a large tumor.

The Urinary System

The cells of the body produce an enormous amount of waste. Some of it, like carbon dioxide, is removed by the respiratory system. Other wastes are removed by the liver and skin. Much of the rest is removed by the urinary system, the topic of this module. In the process of removing wastes, the urinary system also helps maintain internal conditions vital for our health. We'll explore the urinary system in this chapter, looking at its structure and function and a few of the most common diseases as well.

The Urinary System

Of all the organs that remove waste from our bodies, two organs of the urinary system—the kidneys—rank as the most important. The kidneys rid the body of the greatest variety of dissolved wastes. As FIGURE 8-1A shows, each of us has two kidneys. The kidneys don't work alone, however. They're aided by several other organs. Together, these organs form the **urinary system**. Let's take a quick look at each.

The kidneys lie on either side of the backbone, the vertebral column. About the size of a person's fist, the kidneys are surrounded by a cushioning layer of fat.

The human kidneys are oval structures, slightly indented on one side, much like kidney beans (FIGURE 8-1B). (Kidney beans get their name from the fact that they're shaped like this excretory organ.) As illustrated in FIGURE 8-1, each kidney is served by a large artery that branches from the abdominal aorta. The aorta is a large blood vessel that delivers blood to the abdominal organs and lower limbs. Blood drains from the kidneys via large veins that empty into the inferior vena cava.

Waste products are extracted from the blood by numerous microscope filters inside the kidney. Each of these filtration units is called a **nephron** (NEFF-ron). The wastes removed from the blood are then eliminated in the **urine** (YUR-in), a yellowish watery fluid produced by the kidneys.

Urine is drained from the kidney by two tubes, the **ureters**. They extend downward to the urinary bladder. The **urinary bladder** lies in the pelvic cavity just behind the pubic bone. With walls made of smooth muscle that stretch as the organ fills, the bladder

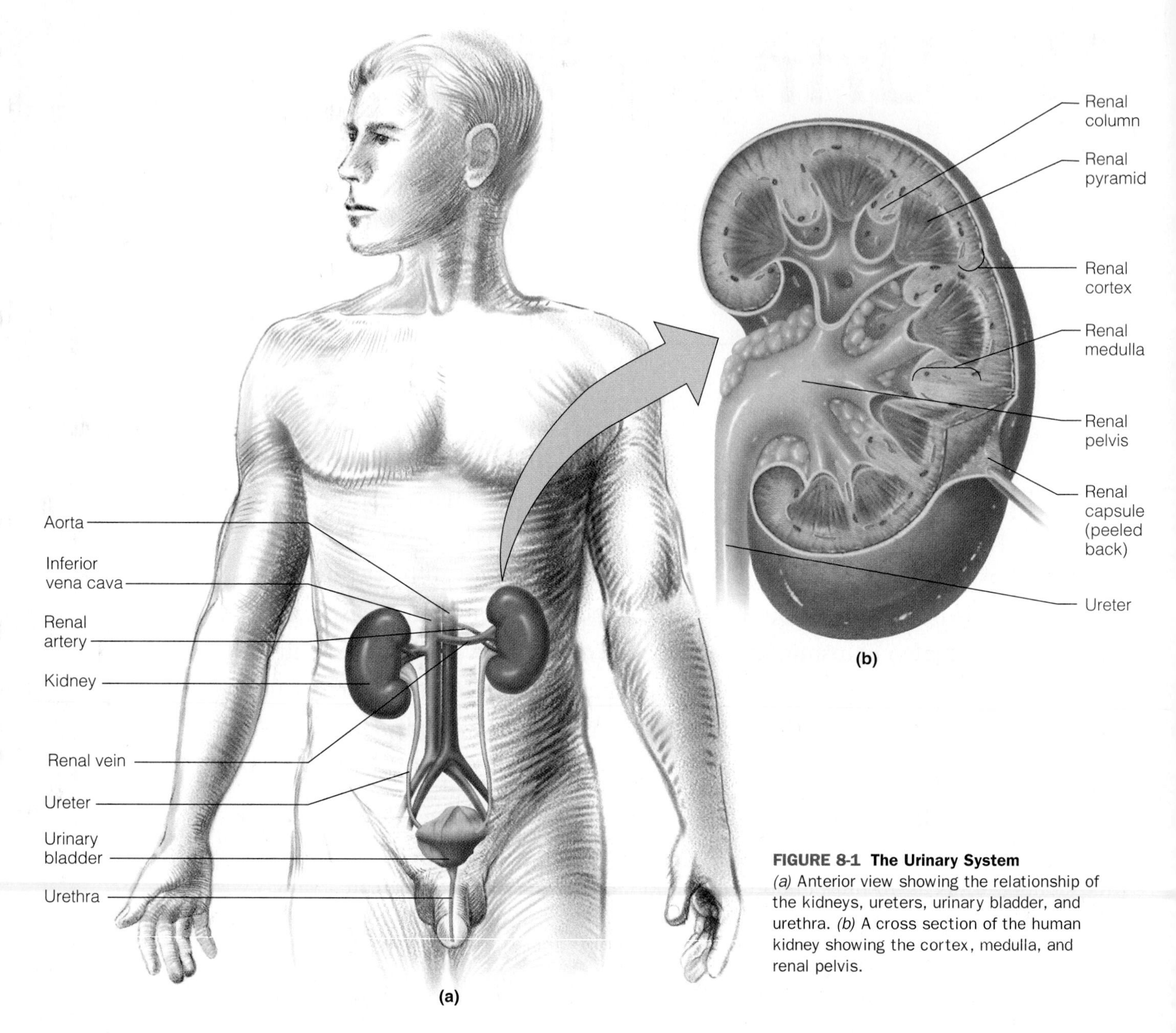

FIGURE 8-1 The Urinary System
(a) Anterior view showing the relationship of the kidneys, ureters, urinary bladder, and urethra. *(b)* A cross section of the human kidney showing the cortex, medulla, and renal pelvis.

serves as a temporary receptacle for urine. When full, the bladder's walls contract, forcing the urine out through the urethra.

The **urethra** is a narrow tube, measuring approximately 4 centimeters (1.5 inches) in women and 15-20 centimeters (6-8 inches) in men. The additional length in men largely results from the fact that the urethra travels through the penis (**FIGURE 8-2**).

The difference in the length of the urethra between men and women has important medical implications. The shorter urethra in women, for example, makes them more susceptible to bacterial infections of the urinary bladder. Bladder infections may result in an itching or burning sensation and an increase in the frequency of urination. Infections may also cause blood to appear in the urine. Urinary tract infections can be treated with antibiotics. Untreated infections may spread up the ureters to the kidneys where they can seriously damage the organ.

Structure and Function of the Nephron

Each kidney contains 1-2 million nephrons. Each nephron, in turn, consists of a tuft of capillaries, known as a **glomerulus** (glom-ERR-you-luss), and a long, twisted tube, the **renal tubule** (**FIGURE 8-3**).

The nephrons filter enormous amounts of blood and produce from 1–3 liters of urine per day, depending on how much fluid a person consumes. Here's how the process occurs. First, arterial blood flows into the glomerulus. Here, waste and even valuable nutrients are filtered out of the blood. The waste-filled fluid then enters the renal tubule.

Each day, approximately 180 liters (45 gallons) of liquid is removed from the blood by the glomeruli. However, as just noted, the kidneys produce only about 1-3 liters of urine each day. Thus, only about 1% of the filtered material actually leaves the kidneys as urine. What happens to the rest of the fluid?

Most of it is reabsorbed. That is, most of it passes from the renal tubule back into the bloodstream, conserving water, valuable nutrients, and ions that were filtered out along with the waste. These materials pass out of the renal tubule into the capillaries that surround each nephron.

This process, called *tubular reabsorption*, is valuable because it conserves water and important ions and nutrients filtered out in the glomerulus. Waste products, such as urea produced by the breakdown of amino acids, however, remain in the fluid inside the renal tubule. They will eventually be excreted in the urine.

Urine flows out of the nephron into another set of ducts, the collecting tubules. The collecting tubules join to form larger ducts that empty their contents into a central chamber inside the kidney. It flows from there into the ureters.

Urination

Urine is produced continuously by the kidneys. It flows down the ureters to the urinary bladder where it is stored until excreted. Leakage out the bladder and into the urethra is prevented by two sphincters—muscular "valves"(Figure 8-2).

The first sphincter, the **internal sphincter**, is formed by smooth muscle in the neck of the urinary bladder at its junction with the urethra. The second valve, the **external sphincter**, is a flat band of skeletal muscle that forms the floor of the pelvic cavity. When both sphincters are relaxed, urine can be expelled.

Urination is stimulated by the accumulation of urine inside the bladder. You need about 200-300 milliliters to start the process. When the bladder fills, the walls of the organ stretch (**FIGURE 8-4**). Special sensors in the walls, called *stretch receptors*, detect the swelling and send nerve impulses via nerves to the spinal cord. These impulses stimulate nerves located in the spinal cord. They, in turn, send impulses back to the smooth muscle cells in the wall of the bladder. These impulses stimulate the smooth muscle cells to contract. In babies and very young

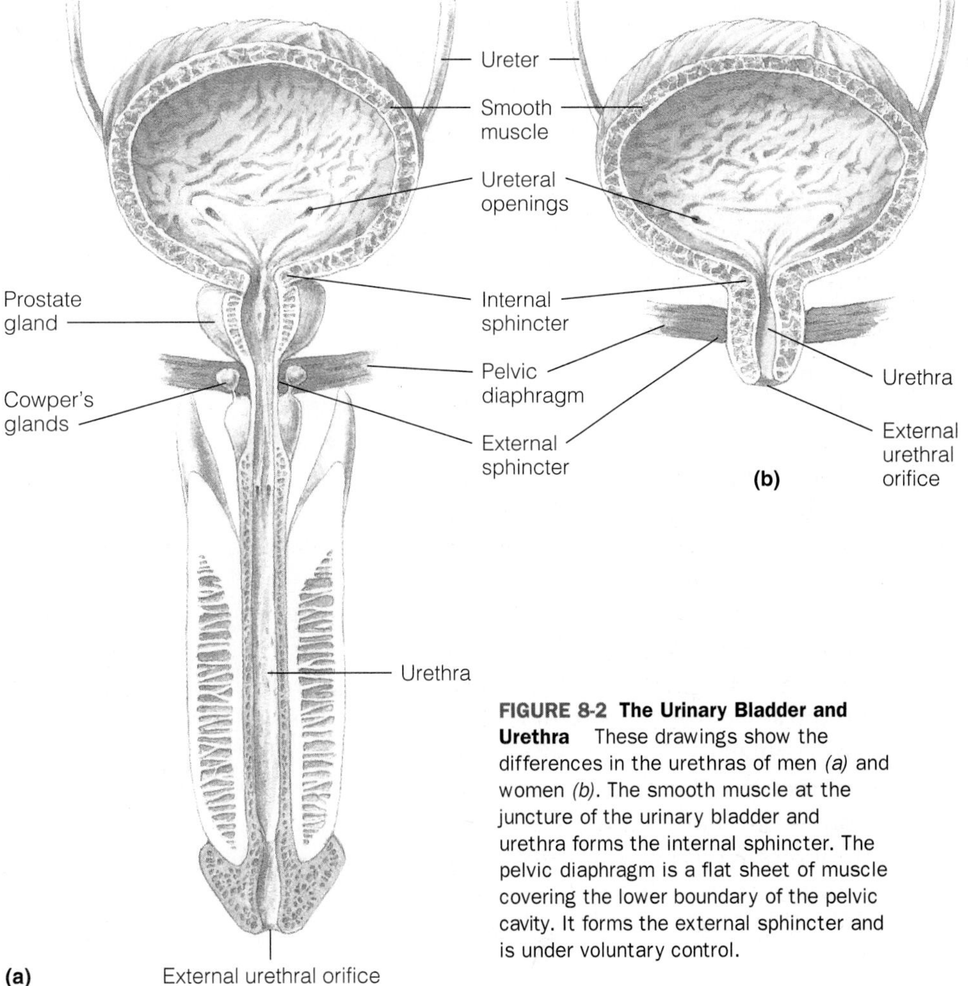

FIGURE 8-2 The Urinary Bladder and Urethra These drawings show the differences in the urethras of men *(a)* and women *(b)*. The smooth muscle at the juncture of the urinary bladder and urethra forms the internal sphincter. The pelvic diaphragm is a flat sheet of muscle covering the lower boundary of the pelvic cavity. It forms the external sphincter and is under voluntary control.

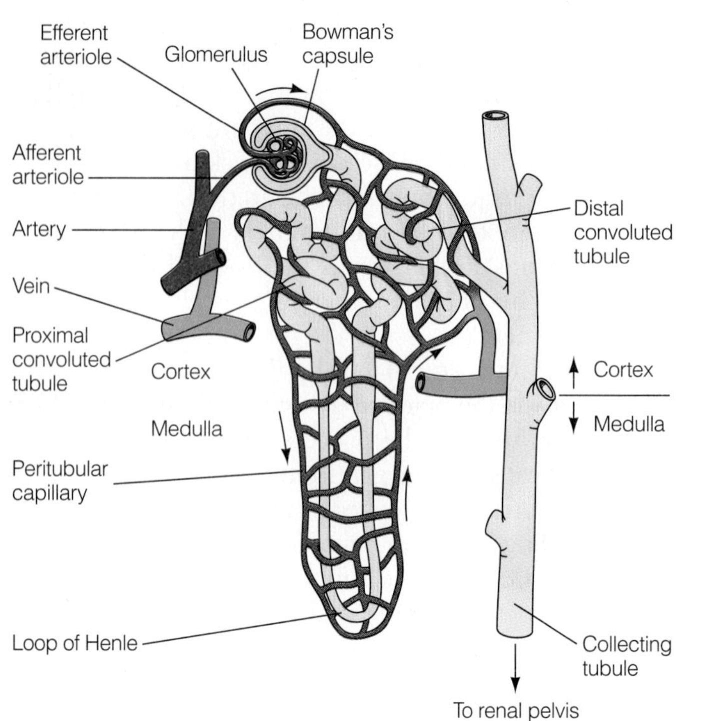

FIGURE 8-3 The Nephron A drawing of a nephron.

children, muscular contraction of the bladder forces the internal and external sphincters open, letting urine enter the urethra. That is to say, the bladder empties entirely by reflex. It is not until children grow older (2-3 years) that they can begin to control urination. It is at this time that the external sphincter comes under conscious control. It will not relax until consciously permitted to do so.

Adults sometimes lose control over urination, resulting in a condition called **urinary incontinence** (in-KAN-teh-nance). Urinary incontinence may be caused by an injury to the spinal cord. This can disrupt signals from the brain that allow us to prevent urination. In such instances, the bladder empties as soon as it reaches a certain size, as it does in babies.

Mild urinary incontinence—the escape of urine when a person sneezes or coughs—is fairly common in women. It usually results from damage to the external sphincter during childbirth. To avoid this, many women exercise the muscles of the pelvis to strengthen them before and after childbirth. Urinary incontinence may also occur in men whose external sphincters have been injured in surgery on the prostate, a gland that surrounds the neck of the urinary bladder.

Controlling Kidney Function

The kidneys help the body control the chemical composition of the blood and tissue fluids by removing wastes. The kidneys also regulate the amount of water that is in our blood. This, in turn, helps to maintain proper concentrations of ions and nutrients essential for normal body function.

As noted earlier, much of the water filtered by the glomeruli is reabsorbed by the renal tubules and returned to the bloodstream. The rate of water reabsorption, however, can be increased or decreased to regulate the concentration of chemicals in our blood. The amount reabsorbed varies with the amount of water we consume. When we have consumed excess amounts, the kidneys reabsorb less. In other words, less water is reabsorbed into the blood. Urine output increases. This process helps rid the body of excess water. When we drink less than we need, the amount reabsorbed increases and urine output falls. These processes help conserve water.

Two hormones play a major role in these important homeostatic processes.

One of them is known as ADH, short for **antidiuretic hormone** (ANN-tie-DIE-yur-eh-tick) or **ADH**. ADH is released by a gland at the base of the brain known as the **pituitary** (peh-TWO-eh-TARE-ee).

To understand how ADH secretion is controlled, let's suppose that you're playing softball in the hot summer sun. The heat makes you sweat, and your

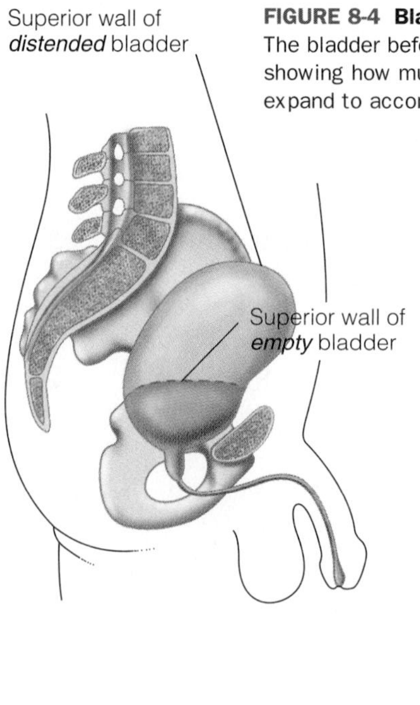

Superior wall of *distended* bladder

Superior wall of *empty* bladder

FIGURE 8-4 Bladder Expansion The bladder before and after it fills, showing how much this organ can expand to accommodate urine.

body begins to lose water. As a result, your blood volume decreases. The concentration of chemicals in the blood, known as the *osmotic concentration*, also increases. The decrease in blood volume is detected by receptors in the heart. The rise in concentration of chemicals in the blood is detected by sensors in the brain. As shown in **FIGURE 8-5**, both of these stimuli trigger the release of ADH from the pituitary gland.

ADH circulates in the blood to the kidney. Here, the hormone stimulates the passage of water from the nephron into the surrounding capillaries. Urine output decreases. Water passing into the bloodstream increases the blood volume and dilutes the chemicals in the blood. When conditions return to normal, the release of ADH stops.

Excess water intake, on the other hand, has just the opposite effect. Drinking more water than you need increases the blood volume and decreases the concentration of chemicals in the blood. The sensors in the brain and heart detect these changes. They then stimulate a reduction in ADH release. As ADH levels in the blood fall, tubular reabsorption decreases. Excess water is removed from the body in the urine. Blood volume and the concentration of chemicals in the blood return to normal.

Water levels are also controlled by the hormone **aldosterone** (al-DOS-ter-own). Aldosterone is a steroid hormone produced by the adrenal glands, which sit atop the kidneys like loose-fitting stocking caps (Figure 8-1). Aldosterone levels in the blood are controlled by blood pressure and blood volume.

Aldosterone release is controlled by a fairly complex sequence of events, too complex to be discussed here. Four our purposes, just remember that aldosterone is released under the same circumstances as ADH. Aldosterone stimulates cells of the renal tubule to increase the amount of sodium and water they reabsorb. This, in turn, increases blood volume and blood pressure.

Water balance is also affected by several chemicals in many people's daily diet, two of the most influential being caffeine and alcohol. Caffeine is found in coffee and nonherbal teas and in many soft drinks. Caffeine increases urine output, causing a person to lose water.

Ethyl alcohol or ethanol is present in beer, wine, wine coolers, and hard liquor. Like caffeine, ethanol increases urine output. It does so by inhibiting the release of ADH. When ADH levels fall, water

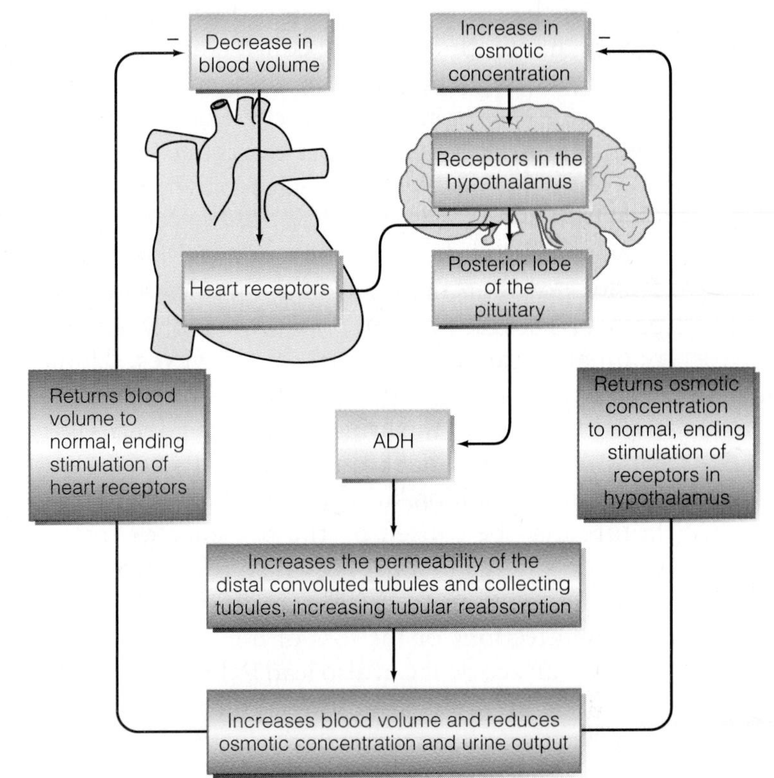

FIGURE 8-5 ADH Secretion ADH secretion is under the control of the hypothalamus. When the osmotic concentration of the blood rises, receptors in the brain detect the change and trigger the release of ADH from the pituitary. Detectors in the heart also respond to changes in blood volume. When volume drops, they send signals to the brain, causing the release of ADH.

reabsorption declines. The result is a marked increase in water loss by the kidneys, supporting the statement that you don't buy wine or beer, you rent it!

Diseases of the Urinary System

Like other body systems, the urinary system can malfunction, creating a homeostatic nightmare in the body. This section discusses two disorders: kidney stones and kidney failure.

Kidney Stones. Urine contains numerous dissolved ions and wastes. When present in excess, higher-than-normal concentrations of calcium, magnesium, and uric acid may crystallize inside the kidney, forming kidney stones (**FIGURE 8-6**).

Kidney stones usually result from inadequate water intake. Most kidney stones are flushed out of the kidney on their own. They enter the ureter and then travel to the urinary bladder. They may then be excreted in the urine. However, the sharp edges of the stone often dig into the walls of the ureters and urethra, causing considerable pain.

Larger stones may become lodged inside the kidney or the ureters and obstruct the flow of urine. This causes internal pressure to increase, which can result in considerable damage to the nephrons if left untreated.

For years, kidney stones were removed surgically. Today, however, there's a much simpler and less painful process. In this technique, physicians bombard kidneys with ultrasound waves. They shatter the stones, producing fine, sandlike grains that are passed in the urine without incident.

Renal Failure. The kidneys may also fail to function, a condition known as **renal failure**. Renal failure may be caused by the presence of toxic chemicals in the blood. It can also result from an immune reaction to certain antibiotics. Severe kidney infections or the loss of huge amounts of blood in an accident can also lead to kidney failure.

Kidney failure is a life-threatening condition, for when the kidneys stop working, water and toxic wastes begin to accumulate in the body. If untreated, a patient will die in 2-3 days.

Patients whose kidneys have shut down, even temporarily, may require **renal dialysis** (DIE-AL-eh-siss). In this procedure, blood is drawn out of a vein and passed through a piece of tubing that carries it to a filter which removes wastes. After filtration, the blood is pumped back into the patient's bloodstream. Dialysis requires several hours and must be repeated every 2 or 3 days. Some patients have dialysis units at home and simply hook themselves up each night before they go to bed.

A simpler and less expensive method is also available. In this procedure, 2 liters of dialysis fluid are injected into a person's abdomen through a permanently implanted tube or catheter (CATH-eh-ter). Waste products diffuse out of the blood vessels into the abdominal cavity. The fluid, containing waste products, is drained from the abdominal cavity a couple of times a day.

This form of dialysis is much simpler, so patients can take care of it themselves. In addition, it allows for more frequent filtering of the blood and permits patients to continue their daily activities without having to strap themselves to a machine.

Complete kidney failure can be treated by kidney transplants. Transplants are generally most successful when they come from closely related family members, for reasons explained in Module 6.

The urinary system is a valuable ally in maintaining homeostasis and good health. Like other organ systems, a healthy lifestyle is vital to maintaining a healthy urinary system.

Kidney stone
Spine
(a) Hip bone
Pelvic cavity

(b)

FIGURE 8-6 Kidney Stones *(a)* An X-ray of a kidney stone. *(b)* Kidney stones removed by surgery. (See nail for size.)

The Nervous System

By now, you've probably gathered that all body systems are crucial for our survival. There is one, however, that plays a dominant role. It is one of two systems that coordinates body functions. It is called the *nervous system*.

Organization of the Nervous System

The human nervous system consists of the brain, spinal cord, and nerves (FIGURE 9-1). This fascinating assembly of nervous tissue governs the functions of the body, exerting control over muscles, glands, and organs. It controls vital functions such as heartbeat, breathing, digestion, and urination. It also helps to regulate blood flow as well as the concentration of chemicals in the blood. As such, the nervous system plays a major role in maintaining homeostasis.

The nervous system receives input from a large number of sources in the body. This input helps the nervous system "manage" body functions in much the same way that letters from citizens help elected officials govern society.

The human nervous system provides functions not seen in other animal species. Our brain, for example, allows us to think about and plan for the future. It enables us to generate ideas. It helps us reason—that is, to judge right from wrong, logical from illogical. In addition, the nervous system also allows us to dramatically change our environment to grow crops to feed millions, to cut down forests to provide wood and paper, and to dig deep beneath the Earth's surface to find gold and other valuable minerals.

The Nerve Cell

The human nervous system consists of the brain, spinal cord, and nerves. The brain and spinal cord make up the **central nervous system**. The nerves comprise the **peripheral nervous system**.

The brain, spinal cord, and nerves are all made of nerve cells or neurons and nervous system connective tissue. Neurons generate impulses when stimulated and transmit these impulses from one part of the body to another.

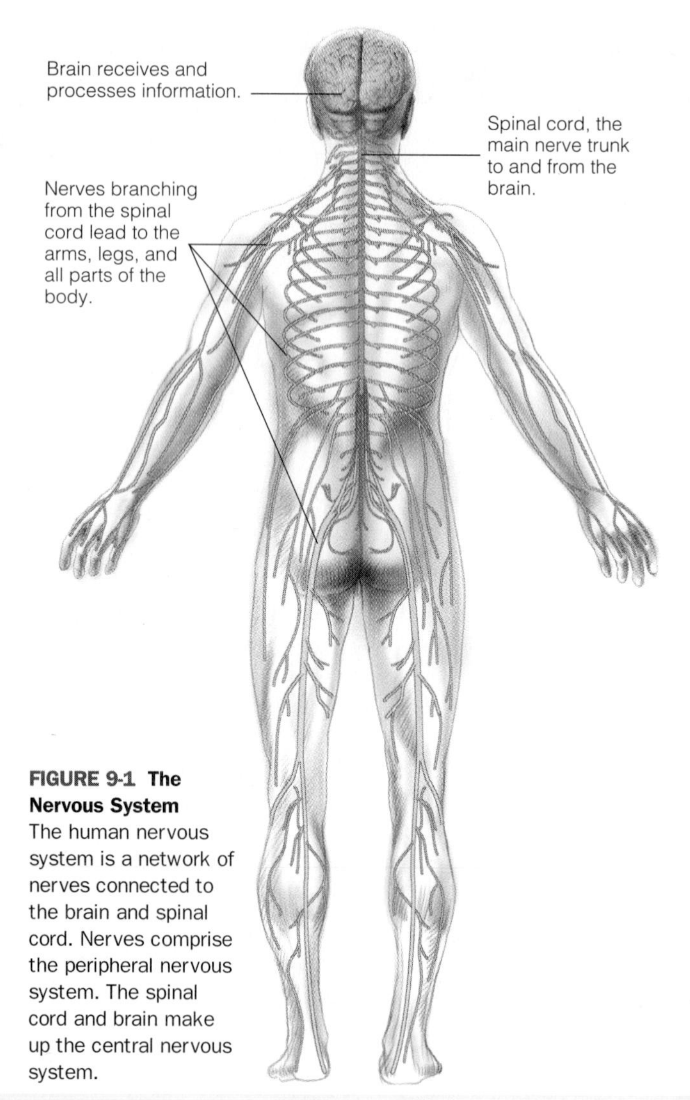

FIGURE 9-1 The Nervous System The human nervous system is a network of nerves connected to the brain and spinal cord. Nerves comprise the peripheral nervous system. The spinal cord and brain make up the central nervous system.

Neurons come in several shapes and sizes. Despite these differences, all neurons consist of a more or less spherical central portion, known as the *cell body* (**FIGURE 9-2**). The cell body houses the nucleus, most of the cell's cytoplasm, and numerous organelles. All nerve cells contain extensions of the cell body that transmit electric impulses. Those that transmit impulses toward the cell body are the **dendrites** (DEN-drights). Those that transmit impulses away from the cell body are known as **axons** (AXE-ons). All nerve cells have multiple dendrites but only one axon.

The most common type of neuron in the human nervous system is the multipolar neuron, so named because it has many dendrites attached to its cell body. The axons of multipolar neurons form nerves in the peripheral nervous system and nerve tracts in the brain and spinal cord.

In both the central and peripheral nervous systems, many axons are coated with a protective layer called the **myelin sheath** (MY-eh-lin) (**FIGURE 9-3**). Myelin is a fatty material laid down in segments by the connective tissue cells of the nervous system. Each segment is separated from the next by a small indentation called a *node*. The myelin sheath is important because it permits nerve impulses to travel with great speed down the axons, "jumping" from node to node (**FIGURE 9-3**).

Although most axons are covered with myelin, some are not. The unmyelinated axons conduct impulses much more slowly. As a rule, the most urgent types of information are transmitted by myelinated fibers; less urgent information is transmitted via unmyelinated fibers.

Nerve cells are highly specialized to perform a very specific function—to conduct signals. Like other highly specialized cells, nerve cells lose their ability to divide and multiply during development. Nerve cells that die, therefore, cannot be replaced like blood cells, liver cells, or skin cells.

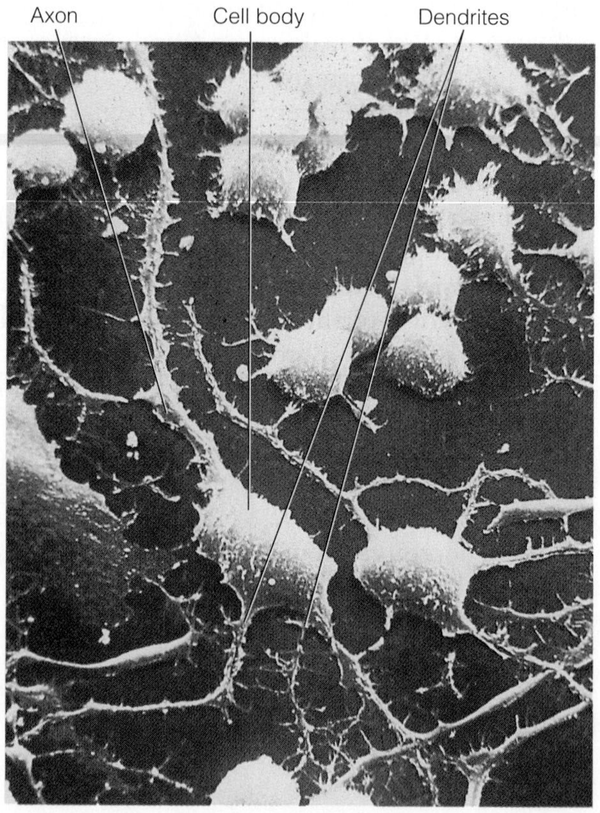

FIGURE 9-2 A Neuron The multipolar neuron resides within the central nervous system. Its cell body has several highly branched dendrites and one long axon. When the axon terminates, it branches many times. Notice the tiny terminal boutons.

Fortunately, not all damage that occurs in nerve cells leads to their death. And some damage can be repaired. For example, if an axon in a nerve in the peripheral nervous system is cut by accident, say by a saw or knife, it may regrow, reestablishing connections with the muscles to which it supplied signals. Partial or nearly complete recovery of control is possible. Neurosurgeons can facilitate regeneration of axons by lining up and sewing the cut ends of nerves together.

As a rule, severed axons in the brain or spinal cord cannot regenerate. Spinal cord injuries that cut nerve tracts, for example, may result in permanent paralysis and a loss of sensation in the region below the damage. The amount of damage depends on where the spinal cord is injured and the severity of the injury.

If the spinal cord injury occurs high in the neck, it severs the nerve fibers traveling to the muscles that control breathing. These muscles become paralyzed, and the person dies quickly. This is how a hangman's noose kills its victim.

Damage to the cord just below the fifth vertebra in the neck does not affect breathing, but it does paralyze the legs and arms. This condition is known as **quadriplegia** (QUAD-reh-PLEE-gee-ah). If the spinal cord injury occurs below the nerves that supply the arms, the result is **paraplegia** (PEAR-ah-PLEE-gee-ah), paralysis of the legs.

Although the outlook for spinal cord injury victims is generally poor, new procedures are being tested that could reduce damage in such instances, even result in restoration of movement and sensation.

Besides being unable to divide, nerve cells require a constant supply of oxygen for energy production. Therefore, if blood flow to the brain is reduced, neurons may begin to die, causing brain damage.

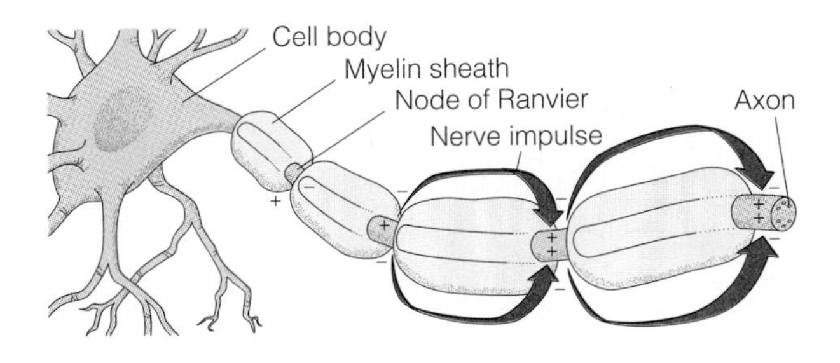

FIGURE 9-3 The Myelin Sheath The myelin sheath allows impulses to "jump" from node to node, greatly accelerating the rate of transmission.

To prevent brain damage from occurring in someone who has collapsed with a severe heart attack or has drowned, rescuers must start resuscitating the victim within 4-5 minutes. Although victims may be revived after this time, the lack of oxygen in the brain often results in brain damage. Generally, the longer the lack of oxygen, the greater the damage.

The only exception is if one drowns in cold water. In such cases, resuscitation may be successful if begun within an hour. Most people recover without any detectable brain damage. The reason for this is that cold water greatly slows brain activity, reducing oxygen demand and cell death.

Nerve cells are also highly dependent on glucose for energy production. As a result, when blood glucose levels fall, nerve cells are first to "feel" the ill effects. Individuals become dizzy and weak. Vision may blur. Speech may become awkward.

What causes nerve impulses? Nerve impulses are not like the electric current that runs through power cords to our computers, which is formed by the flow of electrons. Rather, nerve impulses are caused by the sudden inflow of sodium ions across the cell membrane of nerve cells. After a nerve cell is stimulated, the membrane changes its permeability to sodium ions. Sodium ions flow in, changing the electric charge across the membrane. This, in turn, causes adjacent areas to let more sodium ions in, which changes their charge and so on and so on. Thus, the change in permeability and charge occurring at one point spreads along the membrane. However, stimulated areas quickly return to normal, so another impulse can be transmitted.

In unmyelinated fibers in the human nervous system, nerve impulses travel like waves in water from one region to the next. In myelinated fibers, however, the change in charge "jumps" from one node to the next. This greatly increases the rate at which a nerve impulse travels in myelinated nerve cells, as shown in **FIGURE 9-3**. In fact, a nerve impulse travels along myelinated neuron 400 times faster—nearly 400 miles per hour!

Synaptic Transmission. Several nerve cells are often found in a pathway, so nerve impulses must pass from one neuron to the next. Impulses are transmitted from one neuron to another across a small space that separates them (FIGURE 9-4). This juncture of one neuron with another is called a synapse (SIN-apse).

FIGURE 9-4 The Terminal Bouton and Synaptic Transmission The arrival of the impulse stimulates the release of neurotransmitters held in membrane-bound vesicles in the axon terminals. Neurotransmitters diffuse across the synaptic cleft and bind to the membrane of the next neuron, where it elicits another nerve impulse that travels down the dendrite to the cell body.

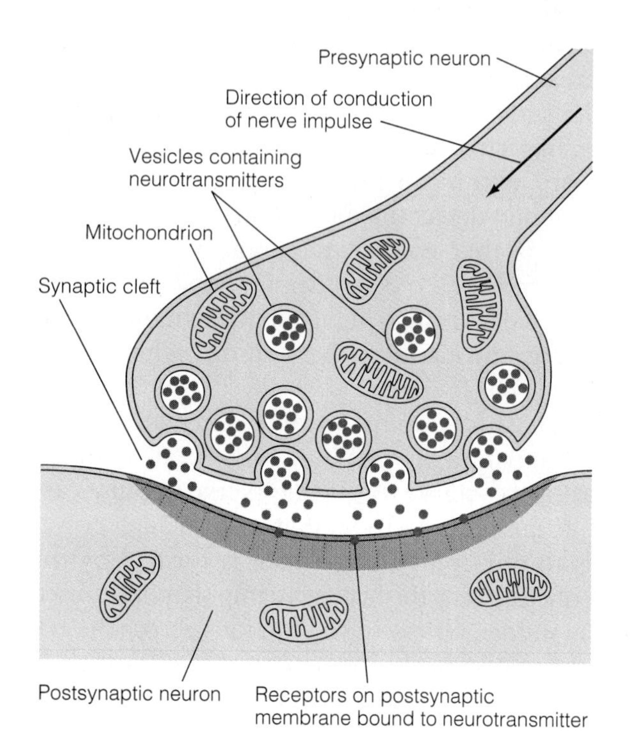

Nerves typically end by branching many times. The ends of the branches form small bulbs called *terminal boutons* (bew-tawns; **bouton** is the French for button). As shown in **FIGURE 9-4**, a small gap exists between the terminal bouton and the cell membrane of the next neuron.

How does a nerve impulse get across the gap? When the nerve impulse ends at the terminal bouton of one neuron, it stimulates the release of a chemical substance within each bouton. These chemicals are known as *neurotransmitters*.

Neurotransmitters are chemical messengers that travel from one nerve cell to the next. They travel across the gap between the nerve cells then bind to receptors in the cell membrane of the next nerve cell. The binding of many neurotransmitters to the next nerve cell can cause a nerve impulse to form. It then moves rapidly down the nerve cell to its target, for example, a muscle.

Transmission across the synapse is remarkably fast, requiring only about 1/1000 of a second, or 1 millisecond. A short burst of neurotransmitter is released each time an impulse reaches the terminal bouton. The neurotransmitter binds to the next nerve cells, causes a change, and then is removed.

Certain drugs and common environmental chemicals such as insecticides impair the removal of neurotransmitters after they've entered the space between nerve cells. Because of this, the neurotransmitters remain bound to the receptors on the nerve cell and the nerve cell continues to fire.

Many insecticides kill insects by disrupting nerve transmission—creating a nervous system overload.

FIGURE 9-5 The Meninges Consisting of three layers, the meninges are the connective tissue covering the brain.

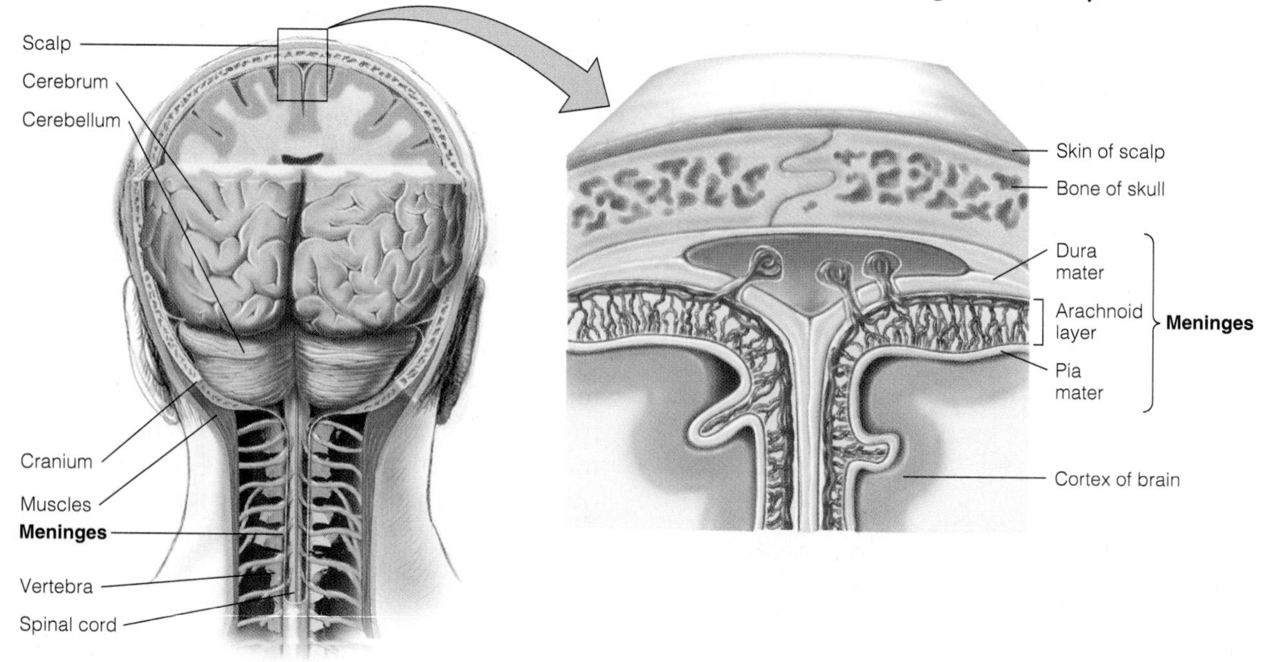

But insecticides can have the same effect on people such as farm workers and pesticide applicators who are exposed to high levels at work. Low doses can cause blurred vision, headaches, rapid pulse, and profuse sweating. Higher doses can be fatal.

Anesthetics (an-es-THET-icks) are chemicals used to deaden pain or to put people to sleep for surgery. Some anesthetics alter the transmission of nerve impulses, decreasing the transmission of pain impulses.

Many antidepressant drugs such as Prozac and Paxil decrease the uptake of a neurotransmitter found in the brain known as *serotonin*. These drugs elevate serotonin levels in the brain that "lifts" depression.

The Structure and Function of the Nervous System

The brain and spinal cord are "housed" in the skull and vertebral canal, respectively, which provide protection against damage. Three layers of connective tissue, known as the **meninges** (men-IN-gees), surround the brain and spinal cord (**FIGURE 9-5**). The space between the middle and innermost layers is filled with a liquid called **cerebrospinal fluid (CSF)** (sir-REE-bro-SPIE-nal). It helps cushion the brain from blows.

The Brain. The brain, shown in **FIGURE 9-6**, consists of several parts. The largest and most conspicuous is the **cerebrum** (sir-EE-brum). The cerebrum is divided into two halves, the right and left **cerebral hemispheres**. Each hemisphere has a thin outer layer of gray matter, containing many nerve cells. This layer is known as the **cerebral cortex** (**FIGURE 9-7**). As shown in Figure 9-7, the cerebral cortex is thrown into numerous folds separated by valleys.

The cerebral hemispheres are divided into four major parts called lobes, shown in Figure 9-6. Each of these lobes carries out specific functions. These areas can be broadly classified into three types: motor cortex, sensory cortex, and association cortex (**FIGURE 9-8**). The **motor cortex** controls muscle activity. The **sensory cortex** receives sensory stimuli such as touch, temperature, taste, and smell. The **association cortex** integrates information, that is, it uses incoming information to bring about coordinated responses.

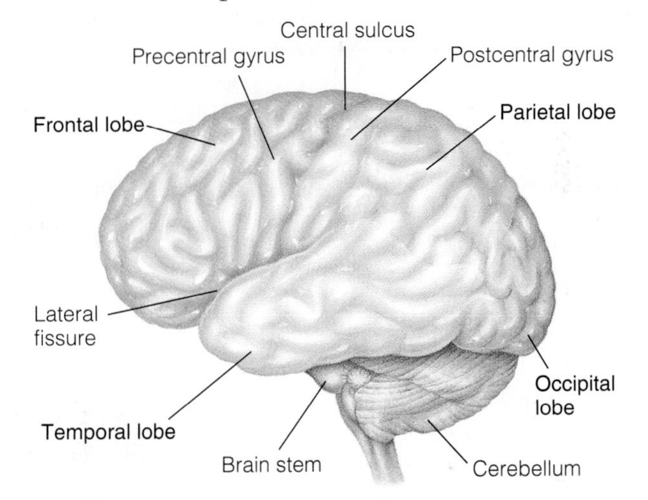

FIGURE 9-6 The Brain The cerebral cortex consists of the lobes shown here. The lobes, in turn, can be divided into sensory, association, and motor areas (not shown).

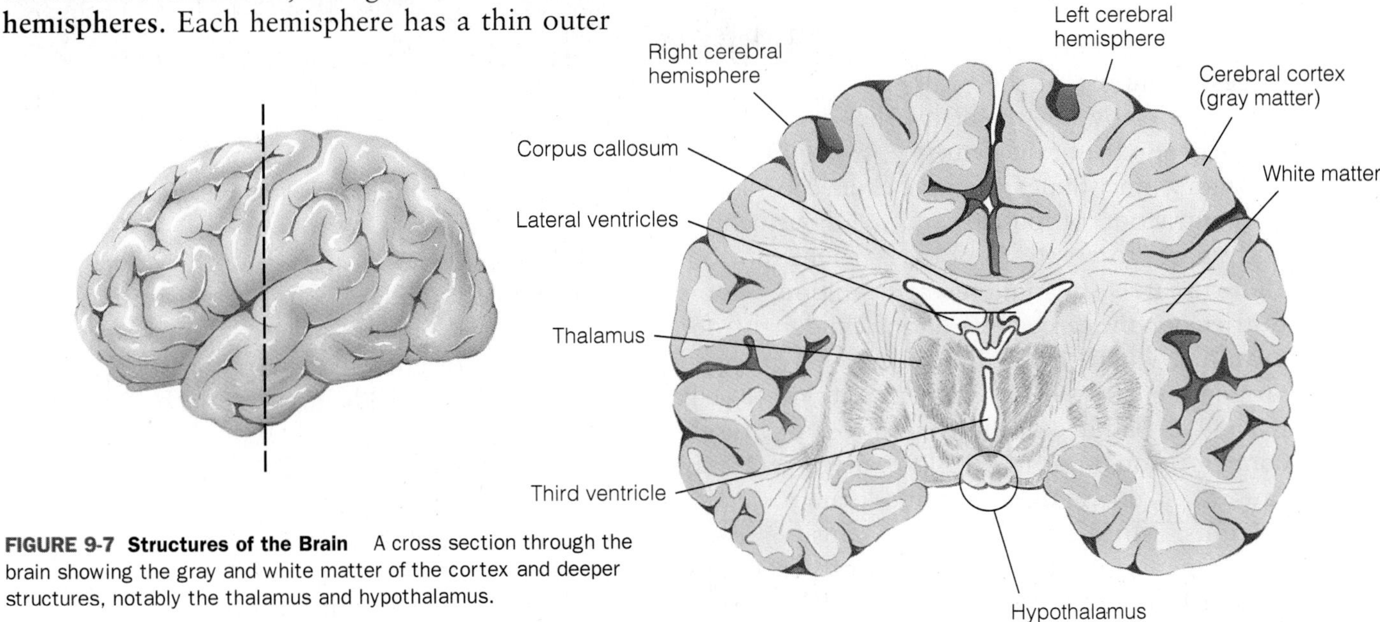

FIGURE 9-7 Structures of the Brain A cross section through the brain showing the gray and white matter of the cortex and deeper structures, notably the thalamus and hypothalamus.

FIGURE 9-8 Functional Regions of the Cortex *(a)* The cerebral cortex has three principal functions: receiving sensory input, integrating sensory information, and generating motor responses. Special sensory areas handle vision, smell, taste, and hearing. *(b)* A pet scan reveals the locations of increased blood flow in the brain during performance of certain tasks.

Key
M. Motor cortex
A. Association cortex
S. Sensory cortex

M. Supplementary motor area (on inner surface—not visible; programming of complex movements)
M. Premotor cortex (coordination of complex movements)
A. Prefrontal association cortex (planning for voluntary activity; decision making; personality traits)
M. Broca's area (speech formation)
S. Primary auditory cortex surrounded by higher-order auditory cortex (hearing)
A. Limbic association cortex (mostly on inner and bottom surface of temporal lobe; motivation and emotion; memory)
M. Primary motor cortex (voluntary movement)
Central sulcus
S. Primary sensory cortex (sensation)
A. Posterior parietal cortex (integration of somatosensory and visual input; important for complex movements)
A. Wernicke's area (speech understanding)
A. Parietal-temporal-occipital association cortex (integration of all sensory input; important in language)
S. Primary visual cortex surrounded by higher-order visual cortex (sight)
(a)

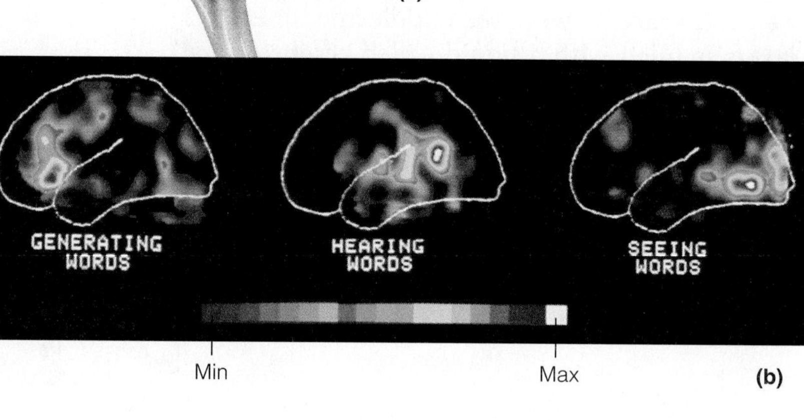

The Cerebellum. Consciousness resides in the cerebral cortex. However, many body functions occur at an unconscious level, among them heartbeat, breathing, and many homeostatic functions. One region of the brain that controls unconscious actions is the **cerebellum** (sar-ah-BELL-um), the second largest structure of the brain. As Figure 9-6 shows, it sits below the cerebrum on the brain stem.

Among other things, the cerebellum coordinates the contraction of muscles and the movement of body parts to create smooth, efficient motion. Unfortunately, in some instances blood flow is restricted to the cerebellum during childbirth—for example, when the umbilical cord is wrapped around the baby's neck. This damages nerve cells in the cerebellum, resulting in a loss of muscle control throughout life. Mild damage generally causes a slight rigidness and moderately "jerky" motions. More severe damage causes serious impairment, with body motions becoming extremely jerky and simple tasks requiring several attempts. This condition is known as **cerebral palsy** (PALL-zee).

Besides ensuring the smooth, coordinated contraction and relaxation of muscles, the cerebellum maintains posture. It receives impulses from sense organs in the ear that detect body position. It then sends impulses to the muscles to maintain or correct posture.

The Thalamus. Just beneath the cerebrum is a region of the brain called the **thalamus** (Figure 9-7). The thalamus is a relay center. It receives all sensory input, except for smell, then relays it to the sensory and association cortex.

The Hypothalamus. Beneath the thalamus is the **hypothalamus** (high-poe-THAL-ah-muss; hypo = under). It consists of many aggregations of nerve cells. These are known as the *nuclei*, not to be mistaken for the nuclei in cells. These groups of cells control a variety of automatic functions, such as appetite, body temperature, and blood pressure.

The Limbic System. Instincts reside in a complex array of structures called the **limbic system** (LIM-bick), shown in **FIGURE 9-9**. The limbic system operates in conjunction with the hypothalamus.

Instincts are among the most fundamental responses of organisms. They include the protective urge of a mother, territoriality, and the fight-or-flight response an animal experiences in the face of danger.

The limbic system also plays a role in emotions—fear, anger, and so on. Electrodes placed in some areas of the limbic system elicit rage. In other areas, they stimulate calmness. Stimulation of specific regions within the limbic system of humans may elicit sensations patients describe as joy, pleasure, fear, or anxiety, depending on the site of stimulation.

The Brain Stem. The brain stem connects the brain to the spinal cord. The brain stem contains aggregations of nerve cells that control many basic body functions such as heart rate, blood pressure, and breathing. It also regulates swallowing, coughing, vomiting, and many digestive functions.

FIGURE 9-9 The Limbic System The odd assortment of structures shown in green is the limbic system. The limbic system is the seat of our emotions and instincts, among other functions.

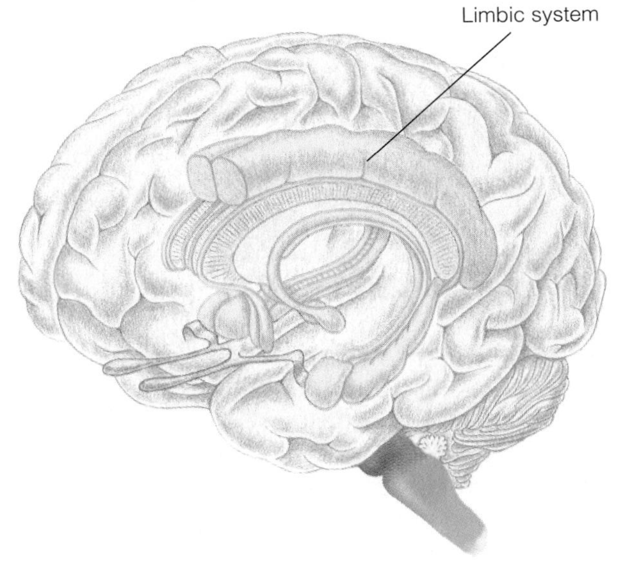

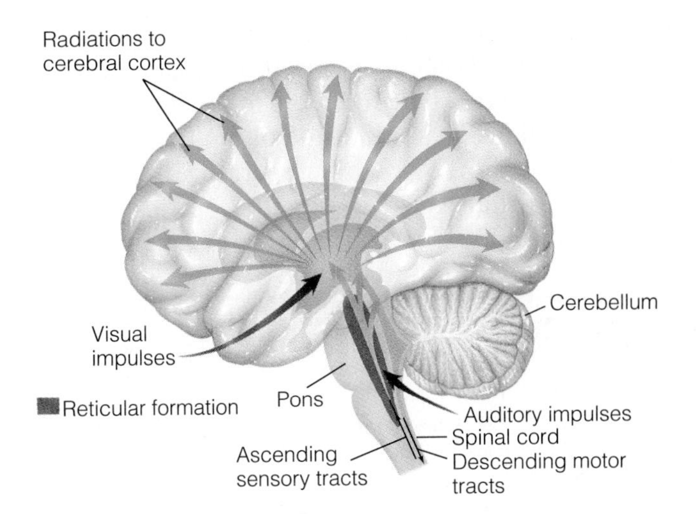

FIGURE 9-10 The Reticular Activating System The reticular formation resides in the brain stem, where it receives input from incoming and outgoing neurons. Fibers projecting from the reticular formation to the cortex constitute the reticular activating system.

The brain stem contains many nerve tracts traveling to and from the brain. Nerve fibers passing through the brain stem, however, frequently give off branches that terminate in a special region of the brain stem known as the *reticular formation* (reh-TICK-you-ler) (**FIGURE 9-10**).

The **reticular formation** runs through the entire brain stem into the thalamus. It receives all incoming and outgoing information. Nerve impulses it receives are transmitted to the cerebral cortex via special nerve fibers. These impulses activate the cerebral cortex. These nerve fibers are like the alarm clock of the central nervous system, helping maintain wakefulness or alertness.

The Spinal Cord. The spinal cord is a long, ropelike structure about the diameter of your little finger. It connects to the brain above and courses downward through a canal formed by the bones of the spine, the vertebrae (ver-tah-BREE) (**FIGURE 9-11**). The spinal cord gives off nerves along its course that supply the skin, muscles, bones, and organs of the body. The spinal cord ends at the lower back, at which point it gives off a series of nerves that supply the lower sections of the body.

Nerves also arise from the brain stem. The nerves that arise from the brain stem and spinal cord form the peripheral nervous system. A nerve is a cordlike structure that consist of numerous nerve fibers. Each fiber is part of a nerve cell, or neuron (NER-on). Nerves transport messages to and from the brain and spinal cord.

Nerve impulses traveling to the brain and spinal cord usually come from receptors located on the ends of nerves. These receptors respond to a variety of stimuli inside and outside of the body, such as blood pressure or body temperature. The nerve impulses that they send to the central nervous system keep the nervous system aware of what's happening inside and outside of our bodies.

Nerve impulses traveling from the brain carry signals to the organs, glands, and muscles. These, in turn, help control body movement and a host of other functions such as digestion.

Part of the peripheral nervous system controls voluntary functions, such as movement of body parts. Other parts control functions, such as heart beat and breathing. These and other functions are largely automatic. They occur, for the most part, without conscious control. Automatic control is absolutely essential to survival. Imagine how much more difficult our lives would be if we had to consciously control our breathing and our heart rate.

Besides having automatic controls to regulate breathing and other essential body functions, the body can respond automatically to certain stimuli as part of reflexes. When a doctor taps a rubber hammer on the tendon just below your kneecap, she is testing one of your body's many reflexes. The tapping stimulates stretch receptors in the tendon. These receptors generate nerve impulses that travel to the spinal cord via sensory neurons. In this reflex, each sensory neuron ends directly on a motor neuron, which supplies the muscles of the thigh. Thus, a quick tap on the tendon results in a motor impulse sent to the muscles on the front of the thigh (quadriceps), causing them to contract and the knee to jerk.

Reflexes are mechanisms that protect the body from harm. Touching a hot stove, for example, elicits the withdrawal reflex. Babies come equipped with a number of important reflexes. Rub your finger on the cheek of a newborn, and it immediately turns its head toward your finger. This reflex helps babies find the mother's nipple. Crying is also a reflex. When a baby is hungry, thirsty, wet, or uncomfortable, it cries, a reflex sure to get attention.

FIGURE 9-11 The Spinal Cord The spinal cord extends from the brain to the lower back.

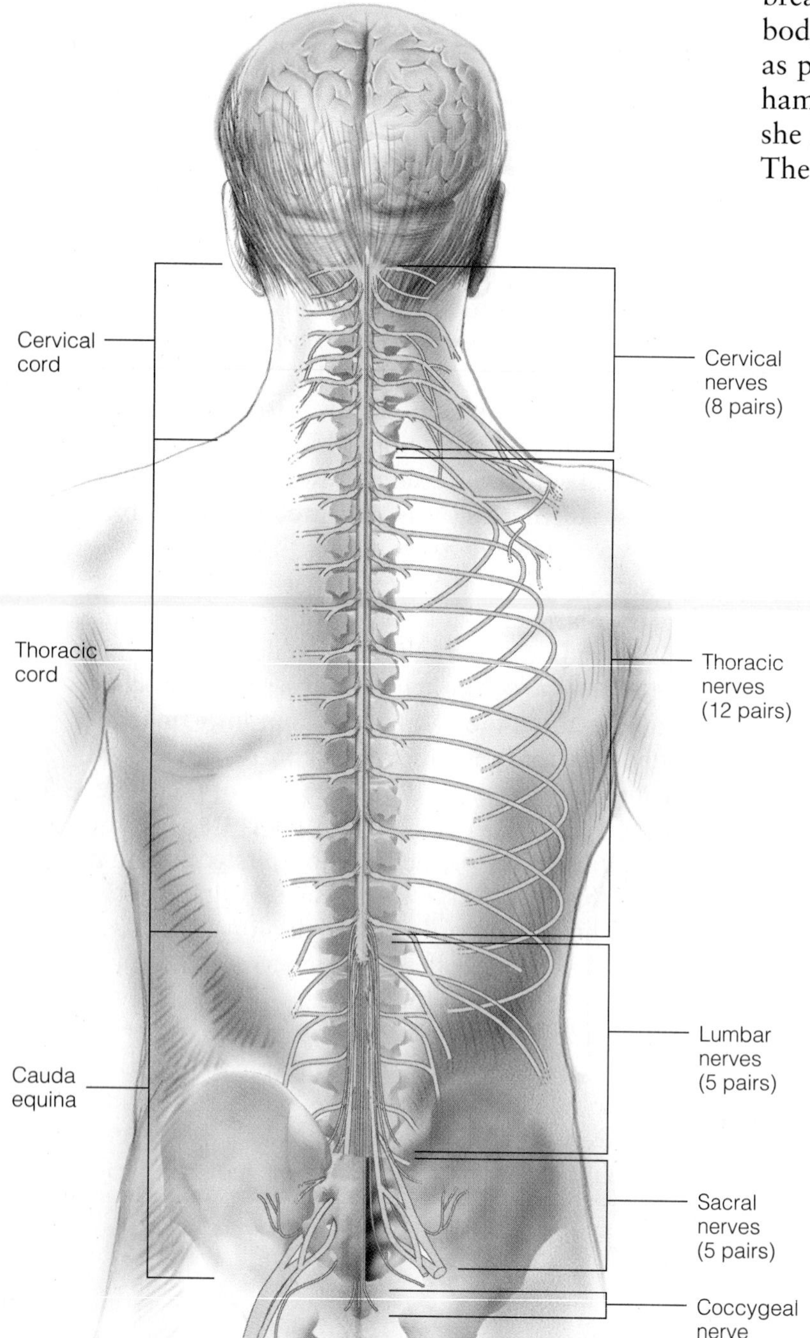

Common Disorders and Diseases

The human brain is subject to a wide variety of disorders from simple headaches to brain tumors to Parkinson's disease. Let's take a look at some of the most common problems.

Headaches. Headaches are the most common form of pain. Many headaches are caused by tension—sustained tightening of the muscles of the head and neck when your are nervous, stressed, or tired. Headaches also commonly result from swelling of the membranes lining the sinuses (cavities inside the bones of the skull surrounding the nasal cavity). Such swelling usually occurs because of sinus infections or allergies. Eyestrain is another common cause of headaches. So is the dilation (expansion) of cerebral blood vessels, which may be associated with high blood pressure or excessive alcohol consumption.

Not all headaches are created equal though. In fact, very serious headaches may result from increased pressure inside the skull. This, in turn, may result from a brain tumor or internal bleeding. Inflammation caused by an infection of the connective tissue coating, the meninges, or the brain itself, also produce intense headaches that require immediate attention.

Less threatening, but still extraordinary, are **migraines**. A migraine headache is a rather severe headache that reappears from time to time. The pain often occurs on one side of the head, and is described as throbbing. It may last from 4–72 hours and is aggravated by movement or physical activity. It is typically accompanied by nausea, vomiting, and sensitivity to light and sound.

Migraines are most common in young adult women. Although a cure is not known, there are treatments. Numerous therapies and medications can be used. Some drugs lessen or eliminate the symptoms; others reduce the frequency or length of a migraine attack.

Like many other neurological disorders, the causes of migraines are not known. Many researchers believe that sudden changes in one's body or environment may trigger migraines in more sensitive individuals.

Multiple Sclerosis. Damage to the myelin sheath of nerve cells in the brain results in a condition known as **multiple sclerosis** or **MS** (skler-OH-siss). The damaged myelin results in nerve cell death that leads to numbness, slurred speech, and paralysis. MS is thought to be an autoimmune disease, that is, a disease in which the immune system attacks the body's own cells.

Early symptoms are generally mild weakness or a tingling or numb feeling in one part of the body. Temporary weakness may cause a person to stumble and fall. Some people report blurred vision, slurred speech, and difficulty controlling urination. In many cases, these symptoms disappear, never to return. Other individuals suffer repeated attacks. Because recovery after each attack is incomplete, patients gradually deteriorate, losing vision and becoming progressively weaker. Fortunately, many treatments are available, and only a small number of multiple sclerosis patients are crippled by the disease.

Stroke. A stroke occurs when blood flow to one part of the brain is halted. It can result when an artery inside the brain breaks, or when an artery is blocked by a blood clot or atherosclerosis (**FIGURE 9-12**). Interrupting blood supply to a part of the brain can result in death, but not always. In some cases, parts of the brain die. Symptoms of a stroke vary, depending on the area that is damaged. However, most stroke victims suffer from a loss of muscle control in one part of the body. A person can recover from a stroke as other parts of the brain take over for the dead cells. Recovery may be partial or complete.

Alzheimer's Disease. Alzheimer's disease is a progressive loss of mental function. It usually begins later in life, mostly in people over the age of 65. Alzheimer's is not life-threatening, but it does result in dramatic change in a person's life. Symptoms start out slowly. Forgetfulness is often one of the most common complaints. Irritability and lack of initiative also occur early on. As mental functions decline, patients lose the ability to remember recent events. Over time, losses may be so great that a parent may be unable to recognize his or her own children. Scientists do not know the cause of Alzheimer's, although there is evidence to suggest that previous severe brain injuries could be the cause in some instances.

Parkinson's Disease. Parkinson's disease is caused by a progressive deterioration of certain brain centers, those that control movement, especially semiautomatic movements such as swinging the arms when walking. It generally occurs later in life, usually in people over 60 years of age.

Parkinson's disease is characterized by shaking of the hands or the head (sometimes both). Tremors disappear or decrease, however, when one moves the afflicted part of the body intentionally—for example, when one reaches for an object. If the disease worsens, symptoms worsen and it becomes difficult to write or walk. Speaking may deteriorate, too, as it becomes difficult to move one's mouth and

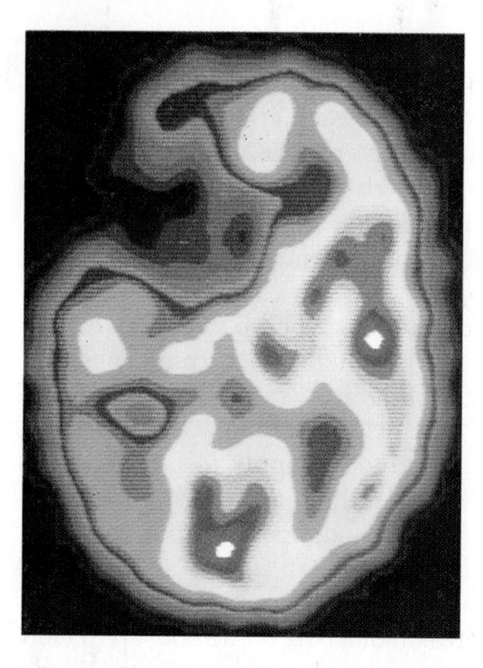

FIGURE 9-12 Brain Damage Caused by Stroke PET (positron emission tomography) scan of brain revealing damaged region of the cerebral cortex following a stroke. This color image was generated by a computer that converts readings of the rate of emission from radioactive glucose molecules injected into the patient. Damaged (dark region) areas show the lowest glucose uptake. Highest glucose uptake and highest emissions are in red.

tongue. Falls become more frequent. It may be difficult to get out of a chair. In later stages of the disease, memory and thinking may deteriorate.

Parkinson's is caused by a lack of dopamine, a chemical substance in the brain. Because of this, patients can be treated with a drug known as *levodopa*. But its benefits may diminish over years of use. And some patients suffer serious side effects, including severe nausea and vomiting. Several other drugs are also used.

Like Alzheimer's, the cause or causes of Parkinson's are unknown. Chemical pollutants may contribute to the disease. Fillings in teeth and chemicals used in wood stains may play a role in the onset of the disease. Very likely, there are many causes.

Recent studies show that fetal cell transplants into the affected brain area may eliminate the symptoms of the disease. A year after receiving a fetal cell transplant, 13 of 19 people showed elevated levels of dopamine production. Surgeons can also destroy a small region of the brain that appears to be hyperactive in Parkinson's patients. Electrodes can also be implanted in the brain to stop uncontrollable movements. Patients are equipped with tiny electronic stimulators, much like pacemakers, which can be switched on or off as needed.

Brain Tumors. Two types of tumors develop in the brain: benign and malignant. Benign tumors are cellular growths that do not grow out of control and do not spread to other parts of the body. Even so, benign tumors can cause problems. They may, for instance, place pressure on areas of the brain resulting in noticeable symptoms. Benign tumors have distinct borders and can be removed surgically; unlike malignant tumors, they do not recur.

Malignant tumors grow rapidly, especially late in the disease. They are likely to place pressure on neighboring structures and may invade into the tissue around them or into other parts of the body.

Brain tumors occur at any age, although the most common groups are children in the 3- to 12-year-old age bracket and adults in the 40- to 70-year-old bracket. Although researchers still do not know what causes brain tumors, there are some risk factors that increase one's likelihood for developing a cancerous tumor. Workers in certain industries, for example, oil refining, rubber manufacturing, and drug manufacturing, are at a higher risk of developing a brain tumor. Chemists and embalmers have a higher incidence of brain tumors as well. Some researchers believe that exposure to certain viruses may be responsible for brain tumors. And some brain tumors seem to run in families, so there may be a genetic cause that is passed from parents to offspring. In most cases, however, patients with brain tumors have no clear risk factors, so researchers believe that the disease is probably the result of several factors acting together.

There's a great deal of concern these days over the potential of cell phones to cause brain cancer. Although several recent studies show no link, cell phone use is a recent phenomenon. Most cancers take 5–30 years to develop.

The brain is a remarkable organ. Working in conjunction with the spinal cord and nerves, it provides a wide variety of services. Most important, it orchestrates many body functions, helping us survive and prosper day-to-day.

The Senses

To survive, all animals—humans included—have to be aware of their environment. We have to know, for example, when danger approaches. We also have to keep track of many other things like changes in temperature and make appropriate adjustments. We must also track conditions inside our bodies. Although much of this tracking is done without conscious awareness, it is essential to our health and survival.

Keeping track of internal and external changes is the responsibility of numerous receptors mentioned in the previous module. Located throughout the body, these receptors are a key part of the nervous system. Strategically located in the skin, internal organs, bones, joints, and muscles, these receptors detect a wide range of stimuli such as pain, temperature, light touch, and pressure. Some even sense body and limb position. The human body also contains highly specialized sensory organs. They detect taste, smell, vision, hearing, and balance.

As you will soon see, some of these receptors are activated by mechanical stimulation—for example, touch or pressure. Others detect chemicals in the food we eat, the air we breathe, or in our blood. Still others are activated by heat and cold. The eye contains receptors that respond to light. And finally, there are receptors stimulated by pinching, tearing, or burning.

The General Senses

Sensors that detect pain, pressure, touch, temperature, and body position send messages to the spinal cord and brain via sensory nerves. These signals may cause a conscious response—for example, the touch of a cat rubbing your leg may cause you to reach down and pat your furry friend. Some stimuli cause unconscious responses—for example,

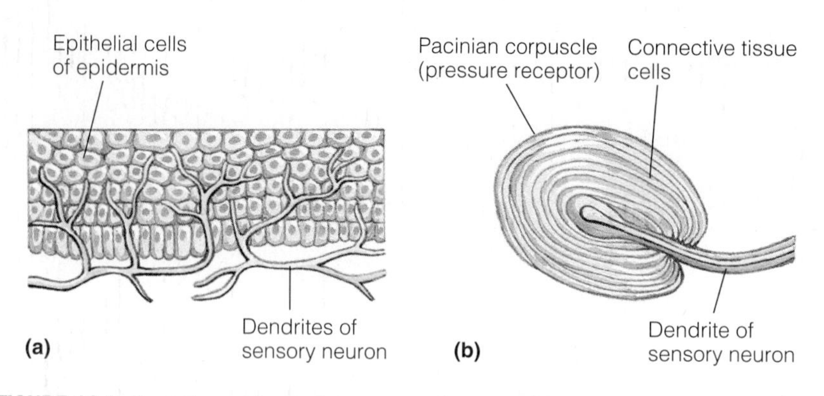

FIGURE 10-1 Receptors General sense receptors are either *(a)* naked nerve endings or *(b)* encapsulated nerve endings.

heat may cause you to perspire. Others may cause no response. They may be simply registered in the brain so you are aware of the stimulus. Finally, some stimuli may be blocked so they elicit no response at all.

Receptors generally fit into two groups based on structure: naked nerve endings and encapsulated receptors (**FIGURE 10-1**). Naked nerve endings are the ends of the dendrites of sensory neurons found in the skin, bones, and internal organs. They are also found in and around joints. They detect pain, light touch, and temperature.

The encapsulated receptors are found in the skin and muscles. One of them is the **Pacinian corpuscle** (pah-sin-e-an core-puss-el), named after its discoverer. It resembles a small cocktail onion pierced by a thin wire, the wire being the nerve ending (**FIGURE 10-1B**). These receptors are stimulated by pressure, such as the pressure you feel sitting in your chair.

Body position is detected by one type of encapsulated receptor in the joints of the body. These receptors inform us of the position of our limbs and alert us to movements of the body. Body position is also detected by two other types of encapsulated receptors found in muscles and tendons. They are stimulated when muscles and tendons stretch.

Taste

In humans, the tongue contains receptors for taste. Known as the **taste buds**, these microscopic, onion-shaped structures are primarily located in the surface of the tongue and on small protrusions, **papillae** (pah-PILL-ee). The papillae are located on the upper surface of the tongue (**FIGURE 10-2A**).

Taste buds are stimulated by chemicals in the food we eat. These chemicals dissolve in the saliva in our mouths, then enter small openings that lead to the interior of the taste bud (**FIGURE 10-2C**). Inside the taste buds are receptor cells that respond to the chemicals.

Humans can discriminate among thousands of tastes. These tastes sensations are a combination of five basic flavors: sweet, sour, bitter, salty, and umami. (Umami is the meaty taste associated with the flavorant MSG (monosodium glutamate)) commonly added to Chinese food.

Smell

The receptors for smell are located in the roof of each nasal cavity. They are found in a patch of cells called the **olfactory membrane** (ole-FACK-tore-ee) (**FIGURE 10-3**). Odors are perceived by special receptor cells, shown in **FIGURE 10-3**. In humans, there are an estimated 50 million olfactory receptor cells in the olfactory membrane. These cells terminate in six to eight long projections, **olfactory hairs**. They contain receptors that bind to chemicals in the air, allowing us to smell.

Humans can distinguish tens of thousands of odors, many at very low levels. Like taste, smell is thought to depend on combinations of primary odors. Unfortunately, scientists studying smell do not agree on what the primary odors are. Some scientists identify seven primary odors ranging from pepperminty to floral to putrid. Various combinations of the primary odors give rise to the many odors we perceive. Other researchers hypothesize that there may be thousands of kinds of smell receptors providing odor discrimination.

The Eye

The human eye is one of the most extraordinary organs in the body. It contains a patch of receptors that respond to light and permit us to perceive the remarkably diverse and colorful environment we live in.

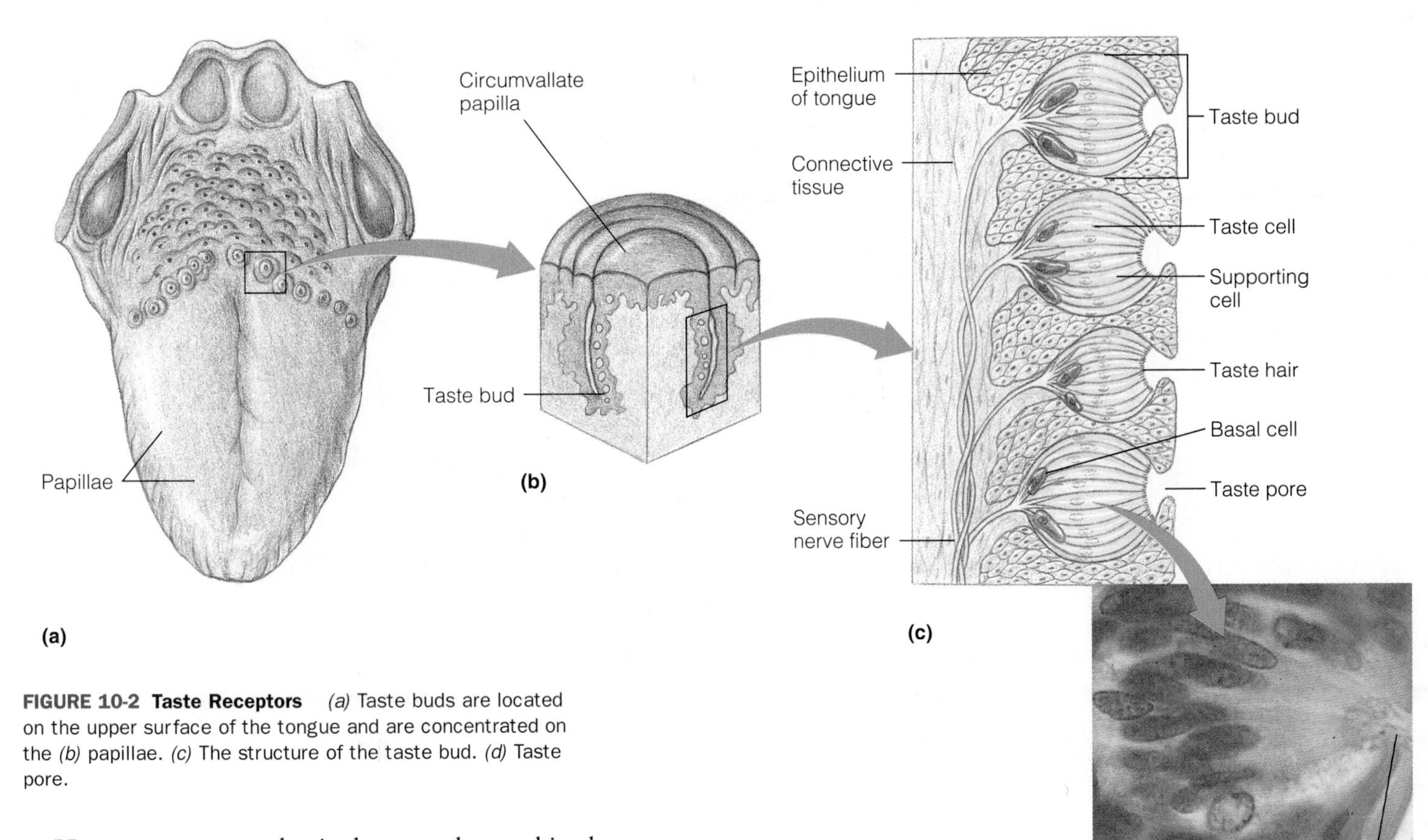

FIGURE 10-2 Taste Receptors *(a)* Taste buds are located on the upper surface of the tongue and are concentrated on the *(b)* papillae. *(c)* The structure of the taste bud. *(d)* Taste pore.

Human eyes are spherical organs located in the **eye sockets** in the skull. The eye is attached to the eye socket by six small muscles that control eye movement.

As FIGURE 10-4 shows, the wall of the human eye consists of three layers. The outermost is a durable, fibrous layer, which consists of the white of the eye. In front, this layer is clear and forms the **cornea.** It lets light into the interior of the eye (FIGURE 10-4).

The middle layer consists of cells containing a large amount of a dark pigment, known as **melanin** (MELL-ah-nin). In front, the pigmented layer forms the iris, the colored portion of the eye visible through the cornea. Looking in a mirror, you can see a dark opening in the iris called the **pupil**. The pupil allows light passing through the cornea to enter the eye. The blackness you see through the pupil is the pigmented region, just mentioned, and the pigmented section of the retina, discussed below.

The iris contains smooth muscle cells. They contract and relax to adjust the amount of light entering the eye. When it is dark, the muscles relax, letting the pupil dilate. More light comes in so we can see better.

The innermost layer of the eye is the **retina**. The retina consists of an outer, pigmented layer and an inner layer consisting of photoreceptors (modified nerve cells that detect light). The retina is weakly attached and can become separated from the middle layer as a result of trauma. A detached retina can lead to blindness if not repaired by surgery. Eye surgeons usually repair detached retinas with lasers.

The **photoreceptors** of the retina are highly modified nerve cells. Two types of photoreceptors are present in the retina: rods and cones. The rods, so named because of their rodlike shape, are sensitive to low light (FIGURES 10-5B and 10-5C). They function at night or in dim light and produce grayish, somewhat indistinct black-and-white images. The **cones**, also named because of their shape, operate only in brighter light. They are responsible for sharp vision as well as color vision.

As FIGURE 10-5B shows, the rods and cones transmit impulses to other neurons in the eye. They, in turn, transmit impulses to another type of neuron. The axons of these cells (ganglion cells) unite to form the **optic nerve**. The optic nerve

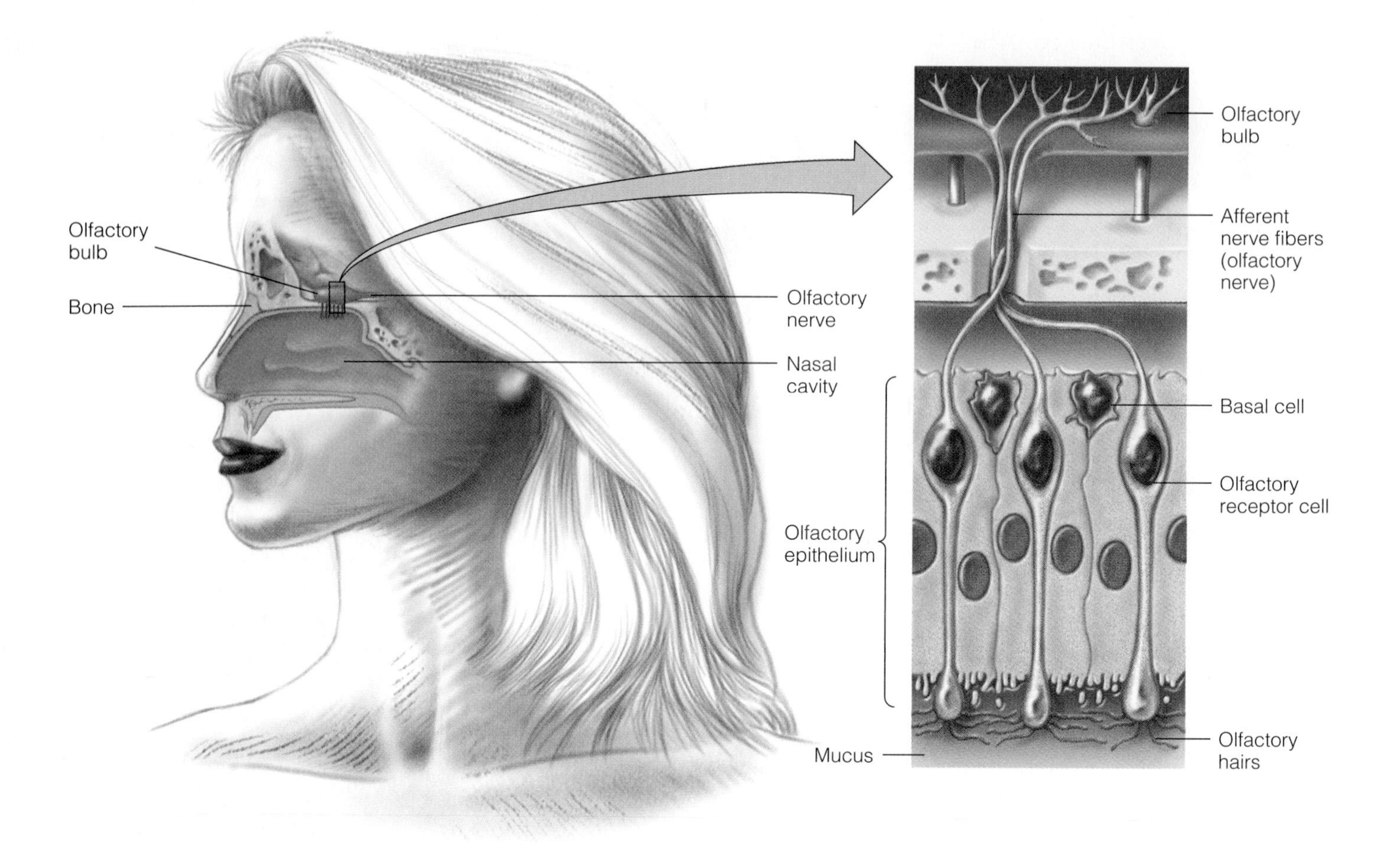

FIGURE 10-3 Location and Structure of the Olfactory Epithelium Odor receptors are located in the olfactory epithelium in the roof of the nasal cavity. Chemicals in the air dissolve in the watery fluid bathing the surface of the cells, then bind to receptors on the plasma membranes of the olfactory hairs. The olfactory receptors terminate in the olfactory bulb. From here, nerve fibers travel to the brain.

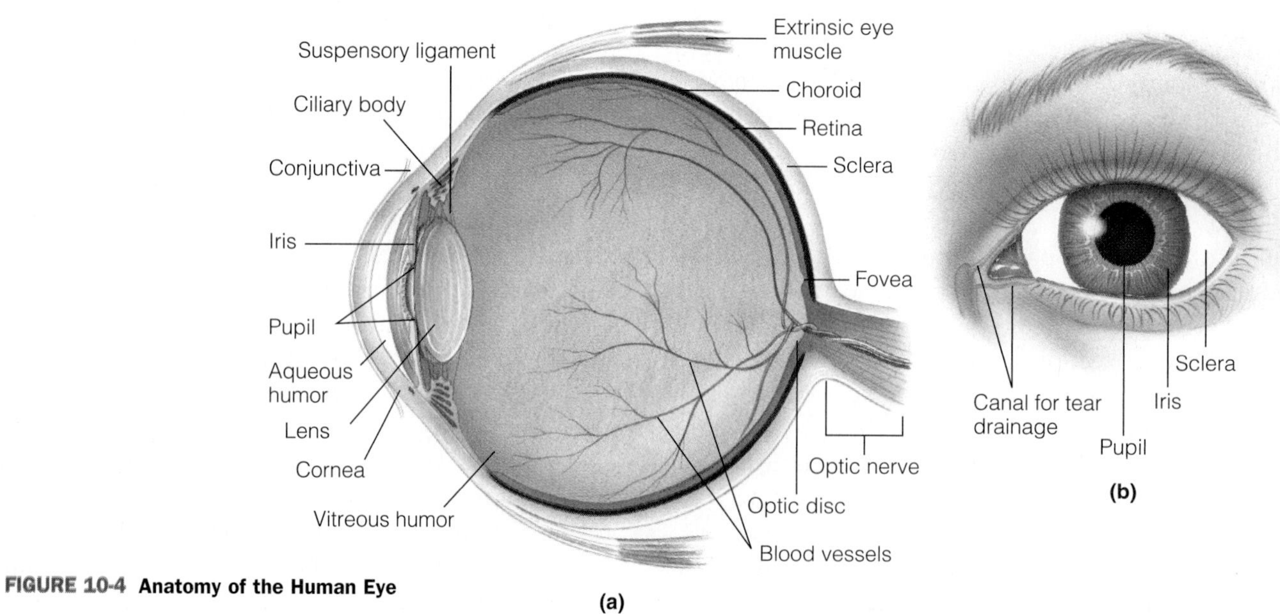

FIGURE 10-4 Anatomy of the Human Eye

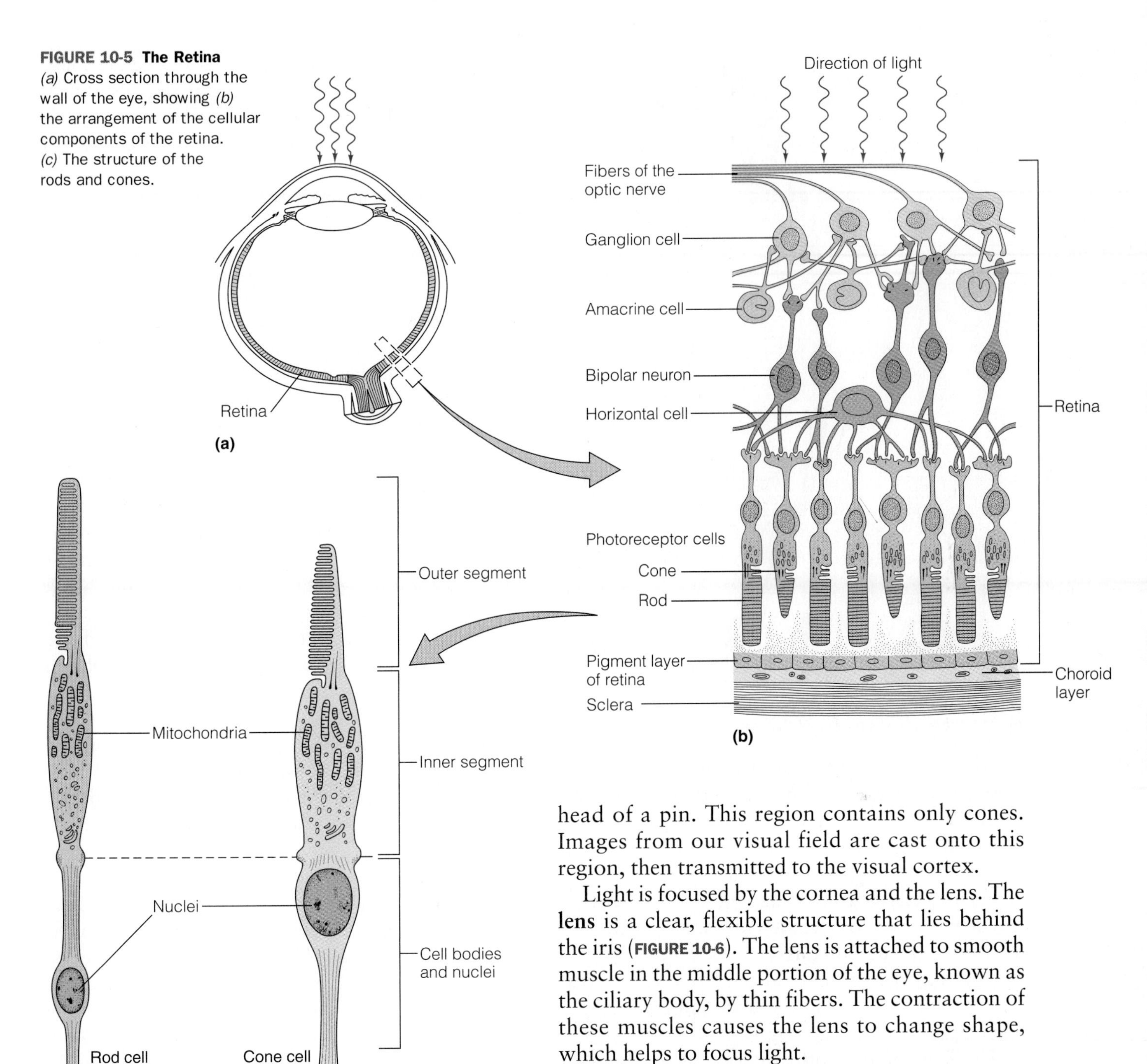

FIGURE 10-5 The Retina *(a)* Cross section through the wall of the eye, showing *(b)* the arrangement of the cellular components of the retina. *(c)* The structure of the rods and cones.

carries nerve impulses to the visual cortex, the part of the brain that receives electrical signals from the retina and produces visual images.

Rods and cones are found throughout the retina, but the cones are most abundant in a tiny region of each eye next to the optic nerve. In the center of this region is a small depression about the size of the head of a pin. This region contains only cones. Images from our visual field are cast onto this region, then transmitted to the visual cortex.

Light is focused by the cornea and the lens. The **lens** is a clear, flexible structure that lies behind the iris (FIGURE 10-6). The lens is attached to smooth muscle in the middle portion of the eye, known as the ciliary body, by thin fibers. The contraction of these muscles causes the lens to change shape, which helps to focus light.

In older individuals, the lens may develop cloudy spots, or **cataracts**. Cataracts are especially prevalent in people exposed to excessive sunlight or excessive ultraviolet light at work or elsewhere. Patients with this disease complain of cloudy vision. Looking out on the world to them is a little like looking through frosted glass. Cataract risk may increase as the Earth's ozone layer, which blocks dangerous ultraviolet radiation from the sun, is eroded by chlorofluorocarbons released from refrigerators, air-conditioning units, and other sources over the past 50 year (Module 17).

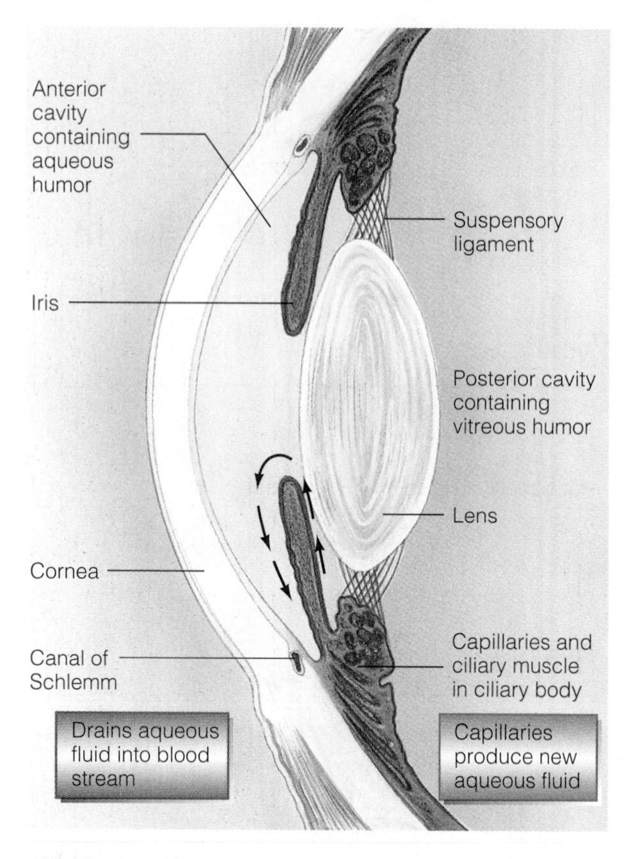

FIGURE 10-6 The Lens Arrows show the flow of fluid through the anterior chamber of the eye.

Dark-eyed people run the highest risk of cataracts. Brown- and hazel-eyed subjects have more cataracts than blue-, gray-, and green-eyed patients. Researchers suggest that melanin in the irises of dark-eyed people may absorb solar radiation, causing more damage to the lens. To lower your risk of cataracts, many eye doctors suggest wearing sunglasses with a coating that reduces ultraviolet penetration. Eye doctors treat cataracts by removing the lens and replacing it with a plastic one.

The lens separates the interior of the eye into two cavities of unequal sizes. Everything behind the lens is the **posterior cavity**. Everything in front of it is the **anterior cavity**.

The posterior cavity is filled with a clear, jelly-like material. The anterior cavity contains a thin watery liquid. In healthy eyes, the production of this liquid is balanced by its absorption into the bloodstream. If the outflow is blocked, however, the fluid builds up inside the anterior chamber, creating internal pressure. This disease is called **glaucoma** (glaw-COE-mah), and progresses very slowly. If untreated, the pressure inside the eye can damage the retina and optic nerve, causing blindness. Because the incidence of glaucoma increases after age 40, doctors recommend an annual eye exam for people over this age. Glaucoma can be treated with eye drops that increase the rate of drainage.

Common Visual Problems

In a relaxed eye with perfect vision, objects farther than 6 meters (20 feet) away fall into perfect focus on the retina (**FIGURE 10-7A**). Many individuals, however, have imperfectly shaped eyeballs or defective lenses. These imperfections result in three visual problems: nearsightedness, farsightedness, and astigmatism.

Nearsightedness or **myopia** (my-OH-pee-ah) results when the eyeball is too long or the lens is too strong (**FIGURE 10-7B**). As a result, light rays arising from distant images fall into focus in front of the retina, creating a fuzzy image. In contrast, nearby images tend to be in focus in the uncorrected eye. Nearsighted people therefore can see nearby objects without corrective lenses—hence the name nearsightedness.

Nearsightedness is quite common. Approximately one of every five Americans needs glasses, contacts, or surgery to correct nearsightedness. Myopia tends to run in families, and it generally appears around age 12, often worsening until a person reaches 20. Myopia can be corrected by contact lenses, glasses, or laser surgery (**FIGURE 10-7B**).

Farsightedness or **Hyperopia** (HIGH-per-OPE-ee-ah) results when the eyeballs are too short or the lens is too weak. In the eyes of farsighted individuals, light rays from distant objects usually fall into focus on the retina, but rays from nearby objects cannot be focused sufficiently. Farsighted individuals, therefore, see distant objects well, but nearby objects are fuzzy. Glasses, contact lenses, or laser surgery can correct this problem (**FIGURE 10-7C**).

Farsightedness is generally present from birth and is usually diagnosed during childhood. Like nearsightedness, it tends to run in families.

The surface of the cornea can also be disfigured. Instead of being rounded like the surface of a basketball, the surface may be shaped more like a football. This condition is called **astigmatism** (a-STIG-mah-tiz-em).

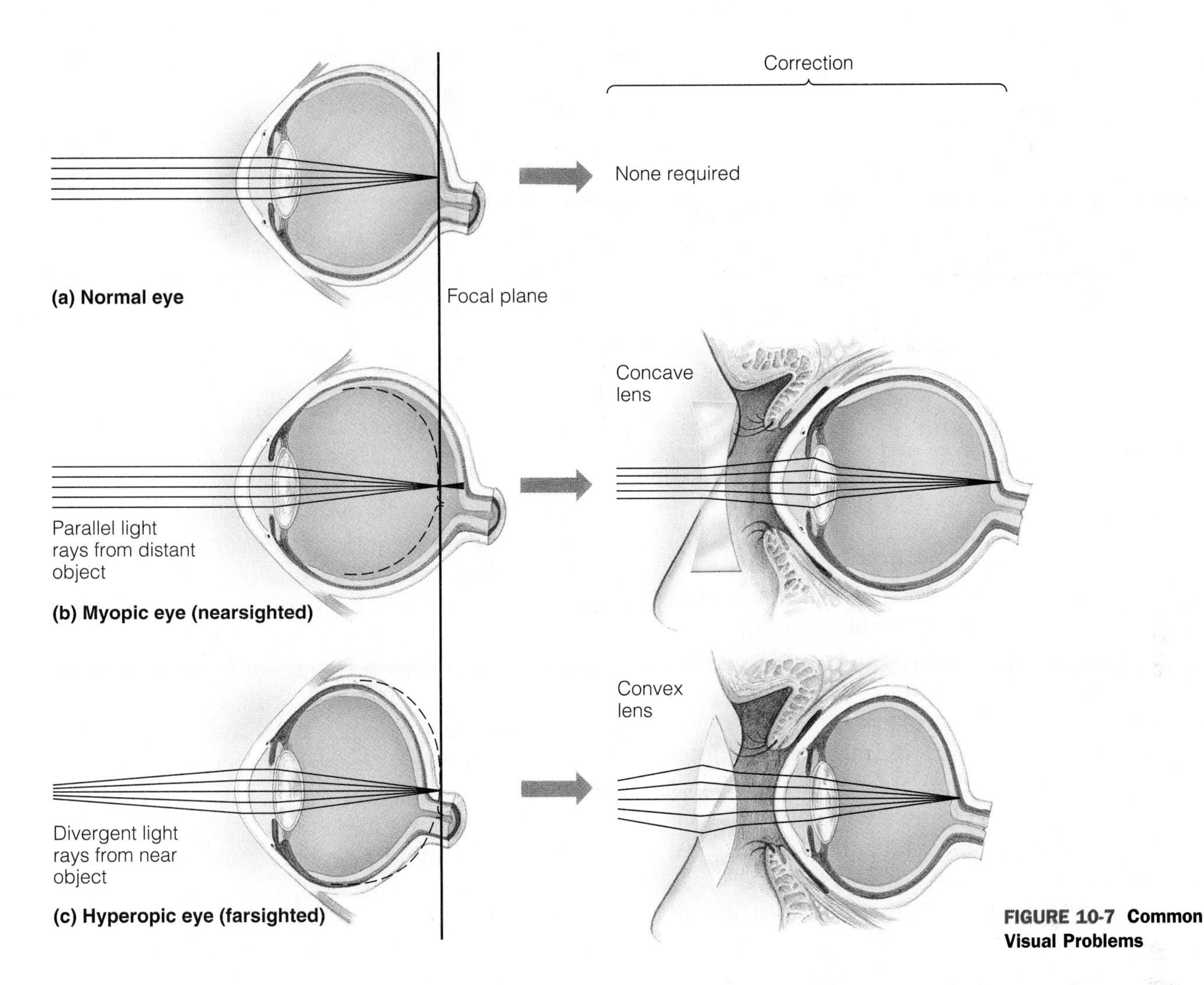

FIGURE 10-7 Common Visual Problems

Astigmatism creates fuzzy images. This condition is usually present from birth and does not grow worse with age. It can also be corrected with glasses, contact lenses, and laser surgery.

Laser Surgery

Eye surgeons have developed a method to correct all three visual problems using lasers to reshape the surface of the cornea. Two types of laser surgery are available in the United States and Canada: PRK and LASIK.

PRK stands for photorefractive keratectomy. In PRK, the surface of the eye is mapped with a computer. It then calculates the amount of tissue that must be removed to correct vision. Another computer then controls laser impulses that shave off the corneal cells correcting the defect.

LASIK stands for the tongue twisting scientific description of the process, *laser in situ keratomileusis.* (Care-ah-toe-mill-E-ew-siss). In this procedure, a computer is used to map the eye to determine the amount of tissue to be removed. However, in LASIK surgery a thin flap of cornea is partially peeled off using an extremely sharp blade, creating a partial flap (**FIGURE 10-8A**). A laser is then used to remove tissue and reshape the cornea (**FIGURE 10-8B**). The flap is then folded back into its original position (**FIGURE 10-8C**). It heals shortly after surgery.

Laser eye surgery is quick and accurate but can cost anywhere from around $500 to $2000 per eye, depending on the level of correction and the surgeon. In addition, the procedure appears to be successful with 95% of patients achieving 20/40 visual acuity or better. Worldwide, hundreds of thousands of patients have been successfully treated with this procedure. Problems arise in under 1% of the cases.

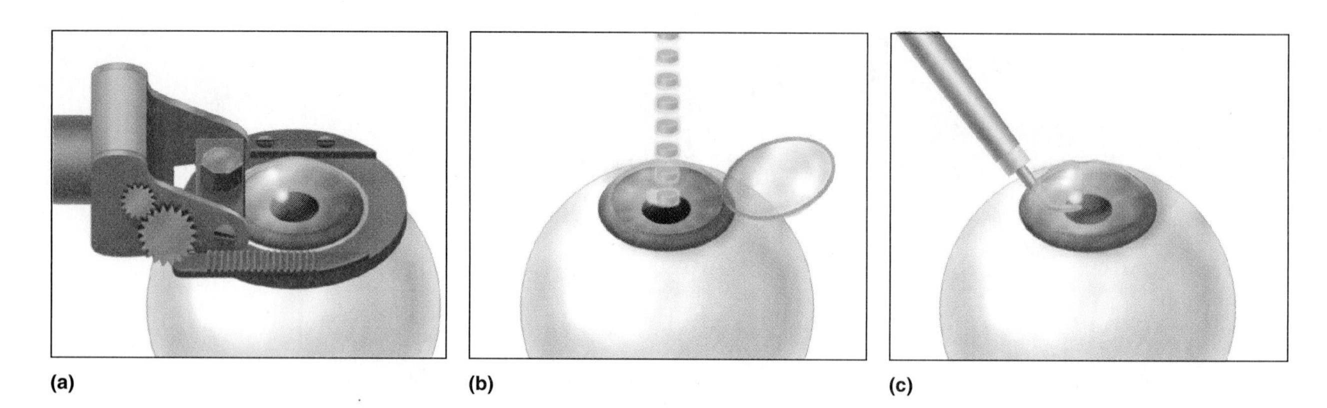

FIGURE 10-8 LASIK Surgery *(a)* Microkeratome slices off a thin layer of corneal tissue *(b)* Laser burns away corneal tissue to correct eyesight. *(c)* Flap is restored, as is vision.

Eyestrain

For most of human evolution, people have used their eyes only for viewing distant objects. Today, however, many of us spend lots of time looking at objects much closer, such as computer screens. This near-point work can strain the eyes and can cause a progressive deterioration of eyesight. Those who read a lot or spend long periods staring at computer monitors may become more nearsighted as they become older. No one knows why, but research suggests that the eye may elongate as a result of constant near-point use.

To reduce eyestrain and the deterioration of eyesight, eye doctors advise that computer operators look away from their screens and that readers look up from their materials regularly, letting their eyes focus on distant objects. This action relaxes the eye muscles, reducing eyestrain.

Presbyopia

As we age, the flexibility of the lens decreases. This makes it difficult for us to focus on close objects. This condition, known as **presbyopia** (PREZ-bee-OPE-ee-a), usually begins around the age of 40. It can be corrected by glasses worn just when reading or working at computers. Presbyopia cannot be corrected by laser eye surgery, because the problem is with the lens, not the cornea.

Color Blindness

About 5% of the human population suffers from color blindness. **Color blindness** is a hereditary disorder and is more common in men than women. Color blindness ranges from an inability to distinguish certain shades of color to a complete inability to perceive color. The most common form of this disorder is red-green color blindness.

In individuals with red-green color blindness, the red or green cones may either be missing or may be reduced in number. If red cones are missing, red objects appear green. If the green cones are missing, green objects appear red. Color blindness can be detected by simple tests (**FIGURE 10-9**).

Many color-blind people are unaware of their condition or untroubled by it. They rely on a variety of visual cues such as differences in intensity to distinguish red and green objects. They also rely on position cues. In vertical traffic lights, for example, the red light is usually at the top of the signal; green is on the bottom. Although the colors may appear more or less the same, the position of the light helps color-blind drivers determine whether to hit the brakes or drive on through.

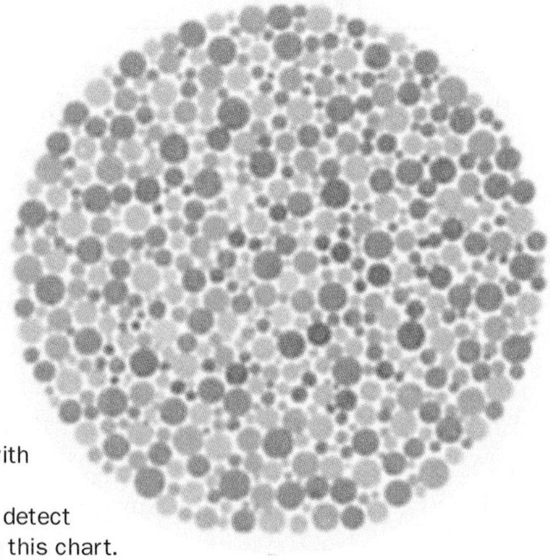

FIGURE 10-9 Color-Blindness Chart People with red-green color blindness cannot detect the number 29 in this chart.

Hearing and Balance

The human ear serves two functions: it detects sound and it detects body position, enabling us to maintain balance. The human ear consists of three parts: the outer ear, middle ear, and inner ear.

The outer ear consists of an irregularly shaped piece of cartilage covered by skin, the **auricle** (OR-eh-kul), and the earlobe, a flap of skin that hangs down from the auricle (**FIGURE 10-10**). The outer ear also consists of a short tube, the **external auditory canal**. It transmits sound waves to the middle ear (**FIGURE 10-10A**).

The middle ear lies entirely within the temporal bone of the skull (**FIGURE 10-10B**). The **eardrum** separates the middle ear cavity from the external auditory canal. The eardrum vibrates when struck by sound waves.

Inside the middle ear are three tiny bones, the hammer, anvil and stirrup, so named because of their shape. The hammer is attached to the eardrum. When the eardrum is struck by sound waves, it vibrates. This causes the hammer to rock back and forth. This, in turn, causes the anvil to vibrate, which causes the stirrup to move in and out. The stirrup attaches to a membrane on the organ of hearing, discussed shortly. When the stirrup vibrates, sound waves are transmitted into this organ.

As Figure 10-10 illustrates, the middle ear cavities open into the back of the mouth via two small tubes, the **auditory tube**. These tubes serve as a pressure-release valve. Normally, the tubes are collapsed. Pressure can build up in the middle ear cavity, for example, when you're on a plane that is taking off. Yawning and swallowing cause the auditory tubes to open, allowing air to flow out of the middle ear cavity. This equalizes the internal and external pressure on the eardrum.

The Cochlea. The inner ear occupies a large cavity in the temporal bone and contains a structure that is shaped like a snail's shell, known as the **cochlea** (COE-klee-ah). It houses the receptors for hearing.

As noted previously, sound waves enter the external auditory canal, where they strike the eardrum. The eardrum vibrates back and forth, causing the bones of the middle ear cavity to vibrate. This transmits sound waves to the cochlea. The cochlea, which houses the receptor for sound, is filled with fluid. Pressure waves inside the cochlea stimulate the sound receptors, called **hair cells** (**FIGURE 10-11B**). When they're stimulated, they send nerve impulses to the auditory cortex of the brain. It translates these impulses into sound.

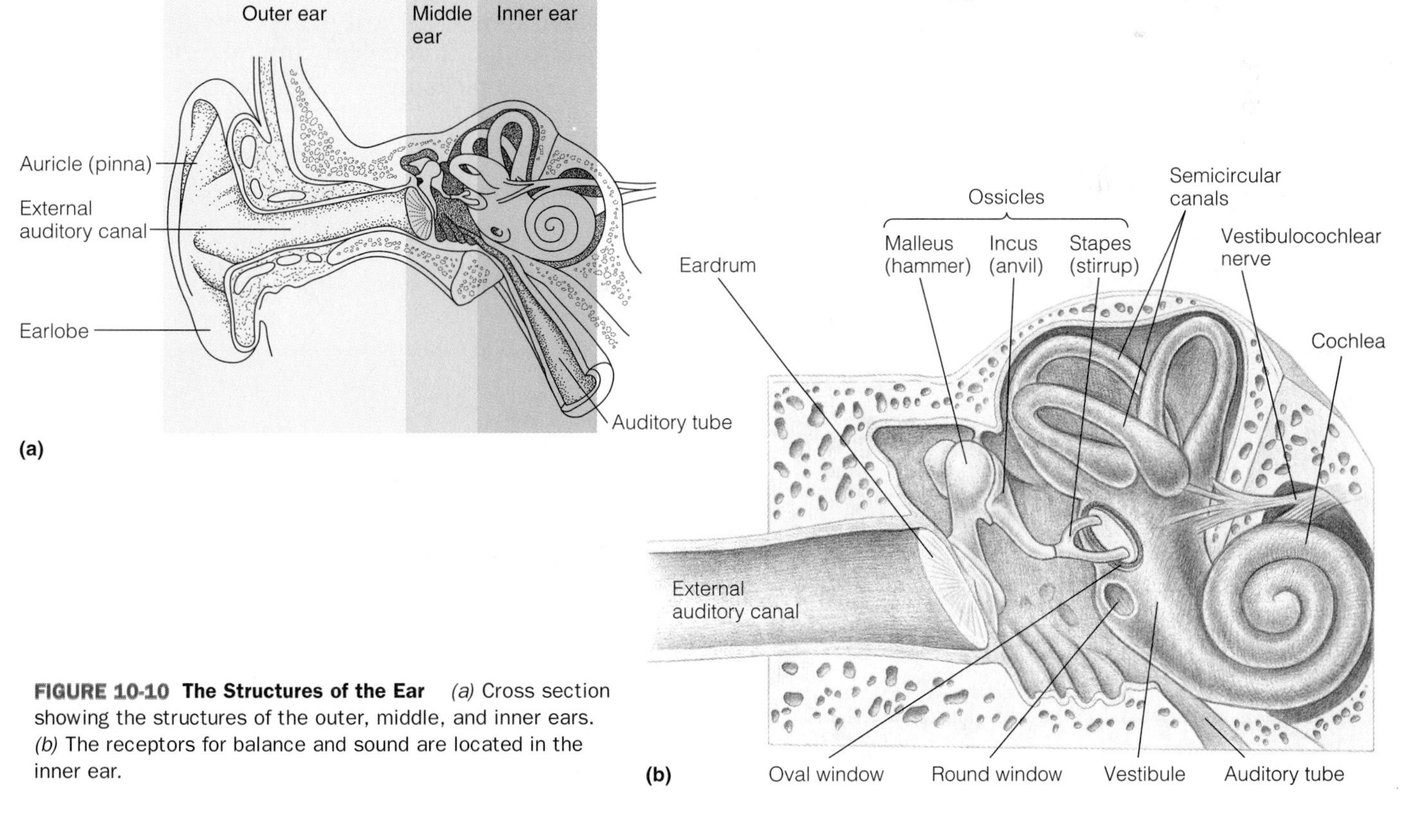

FIGURE 10-10 The Structures of the Ear *(a)* Cross section showing the structures of the outer, middle, and inner ears. *(b)* The receptors for balance and sound are located in the inner ear.

Hearing Loss. As people grow older, many begin to lose their hearing, but hearing loss usually occurs so slowly that most people are unaware of it. In some cases, though, people lose their hearing suddenly. A loud explosion, for example, can damage the hair cells or even break the bones in the middle ear.

Hearing loss falls into one of two categories, depending on the part of the system that is affected. The first is **conduction deafness**. It occurs when the conduction of sound waves to the inner ear is impaired. Conduction deafness most often results from infections in the middle ear. Bacterial infections may result in the buildup of scar tissue that causes the bones in the middle ear cavity to fuse together. They therefore are unable to transmit sound. Infections of the middle ear usually enter through the auditory tube. A sore throat caused by bacteria can spread to the ear, where it requires prompt treatment. Conduction deafness may also result from excessive earwax in the external auditory canal or rupture of the eardrum.

Conduction deafness is treated by hearing aids. A **hearing aid** fits in the ear or just behind it (**FIGURE 10-12**). These devices bypass the defective sound-conduction system by transmitting sound waves through the bone of the skull to the inner ear. These cause fluid pressure waves to form in the cochlea and stimulate the hair cells.

The second type of hearing loss is neurological and is called **nerve deafness**. It may result from physical damage to the hair cells in the cochlea. Explosions, extremely loud noises, and some antibiotics can all damage the hair cells, creating partial to complete deafness.

Damage to the nerve leading from the cochlea to the brain can also cause this type of deafness. The auditory nerve may degenerate, thus ending the flow of information to the cortex.

More than 2 million Americans are profoundly deaf. Until recently, this condition was considered virtually untreatable. Children who are born deaf or are deafened before they begin to speak often fail to mature emotionally. Some profoundly deaf children, in fact, never advance beyond third- or fourth-grade reading levels.

Hearing aids usually cannot help individuals who are born deaf or those who suffer from nerve damage. Researchers, however, have developed a device, called a **cochlear implant** (**FIGURE 10-13**). This device picks up sound and transmits it to a receiver implanted inside the skull. The signal then travels to an electrode implanted in the nerve that carries signals from the sound receptors to the

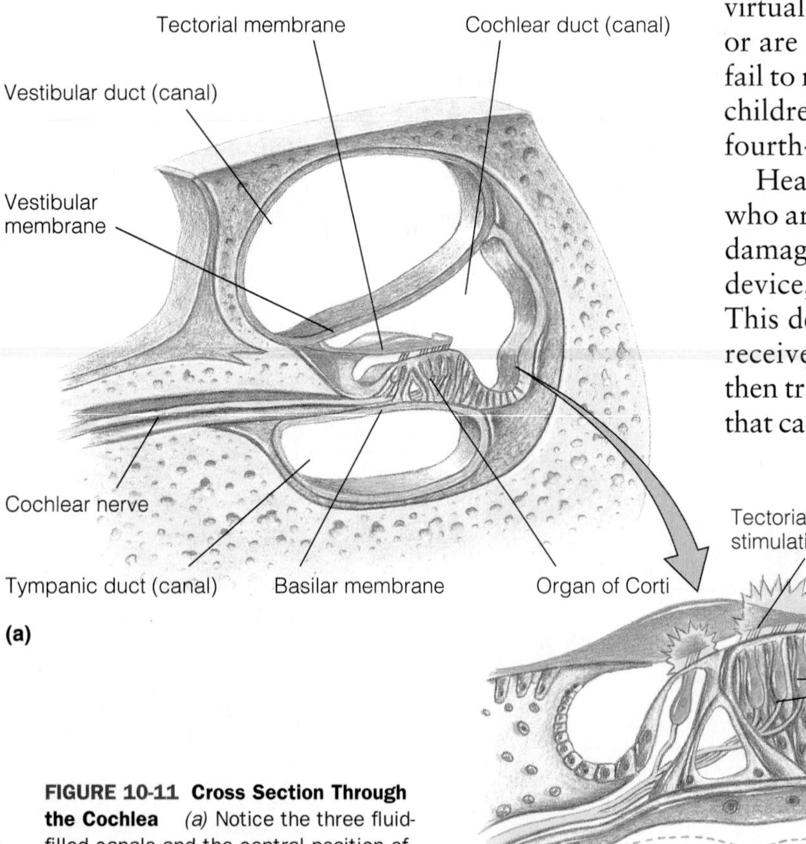

FIGURE 10-11 Cross Section Through the Cochlea *(a)* Notice the three fluid-filled canals and the central position of the organ of Corti. *(b)* Hair cells of the organ of Corti are embedded in the overlying tectorial membrane. When the basilar membrane vibrates, the hair cells are stimulated.

brain. Electrical impulses in the electrode stimulate the nerve. Impulses then travel to the auditory cortex.

Today, hundreds of adults and children are equipped with cochlear implants that detect and transmit a wider range of sounds. Recipients of the new models can perceive many distinct words. Some individuals equipped with these devices have apparently developed a remarkable ability to perceive sounds. Cochlear implants also help the deaf monitor and regulate their voices and make lip reading easier.

The Vestibular Apparatus. The cochlea lies next to the **vestibular apparatus**. It consists of the two parts: the semicircular canals and the vestibule. The vestibule is a bony chamber between the cochlea and semicircular canals. These structures house receptors that detect body position and movement (**FIGURE 10-14**).

The semicircular canals are arranged at right angles to one another. Each canal is filled with a fluid. As **FIGURE 10-14A** shows, the base of each semicircular canal expands. On the inside wall of each of these expanded areas is a receptor that detects movement in the fluid.

Rotation of the head causes the fluid in the semicircular canals to move. The movement stimulates hair cells of the receptors. This creates impulses that are sent to the brain, alerting it to the rotation of the head and body.

FIGURE 10-12 Hearing Aids Worn by people with conduction deafness, hearing aids send sound impulses through the bone of the skull to the cochlea.

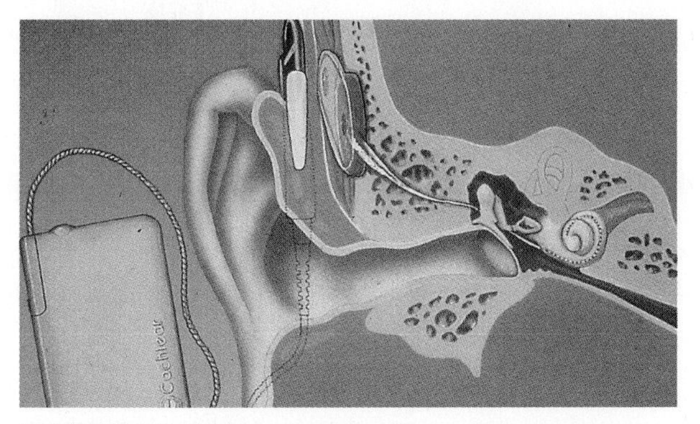

FIGURE 10-13 Cochlear Implant The cochlear implant can correct for nerve deafness. Electrodes convey electrical impulses from a small microphone mounted in the ear to the auditory nerve.

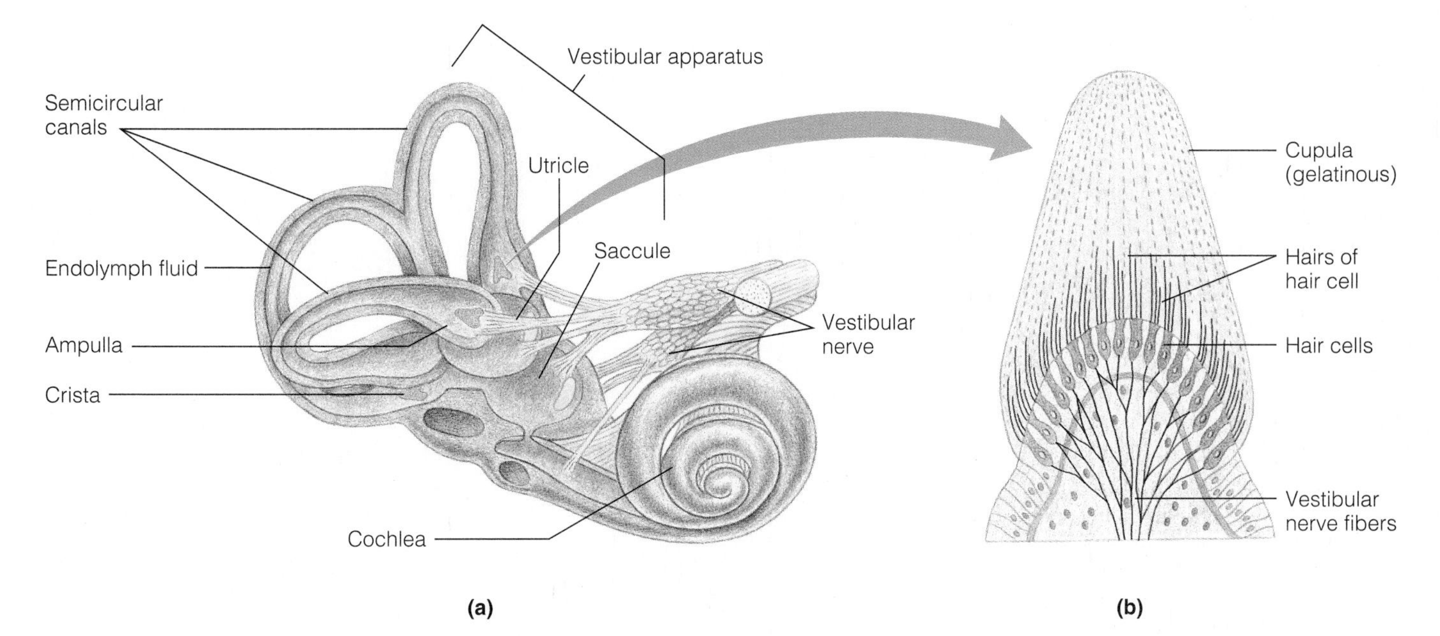

FIGURE 10-14 Vestibular Apparatus and Semicircular Canals *(a)* This illustration shows the location of the receptors in the semicircular canals. (blue) The semicircular canals are filled with fluid. *(b)* When the head spins, the fluid is set into motion, deflecting the gelatinous cupula of the crista, thus stimulating the receptor cells.

Two additional receptors play a role in balance and the detection of movement. They are located in the vestibule. They provide input at rest and thus help us stay balanced when not moving. They also provide information when we are moving in a straight line.

In some people, activation of the receptors in the vestibule may cause motion sickness. It is characterized by dizziness and nausea.

It should be pretty clear by now how important the body's many receptors are. They allow us to work and play and move about in our environment without worry. They're vital for our day-to-day success and our long-term survival.

The Skeletal System

What separates humans and other animals from plants? One of the main differences is that we are animated—that is, we can move about of our own accord. In fact, that's where the term *animal* comes from.

Movement in humans is a complex process involving muscles and our bony skeleton. But movement is also dependent on joints. In this module, we'll explore the skeleton and joints.

Structure and Function of the Human Skeleton

The human skeleton consists of 206 bones (**FIGURE 11-1**). The bones of the skeleton provide internal structural support, giving shape to our bodies and enabling us to maintain an upright posture. Some bones protect internal organs. The rib cage, for instance, protects the lungs and heart, and the skull safeguards the brain. Bones serve as the site of attachment for the tendons of many skeletal muscles whose contraction results in movements.

Inside bones, in the marrow, are the cells that give rise to red blood cells, white blood cells, and platelets. Bones are also a storage depot for fat, needed for cellular energy production at work and at rest. Finally, bones are a reservoir of calcium. Bones release and absorb calcium as needed, which helps maintain normal blood levels. That's important because calcium is required for muscle contraction.

Take a moment to study the skeleton in Figure 11-1. As you examine it, you may notice that bones come in a variety of shapes and sizes. Some are long; some are short; and some are flat and irregularly shaped. Nevertheless, all bones share some characteristics.

First, bones are made of bone tissue. Bone tissue contains numerous bone cells, known as **osteocytes** (OSS-tee-oh-SITES). The bone cells are embedded in a dense material, known as the **matrix** (MAY-tricks). The matrix consists of calcium phosphate, which gives bone its hardness and some of its strength. The rest of bones' strength comes from a special type of protein fiber known as **collagen** (COLL-ah-gin).

As shown in **FIGURE 11-2**, bones typically consist of an outer shell of dense material called **compact bone**. Inside bones is a mass of **spongy bone**. Spongy bone is less dense than compact bone and contains numerous small cavities.

On the outer surface of the compact bone is a layer of connective tissue. It serves as the site of attachment for skeletal muscles. It also contains bone-forming cells that repair broken bones (discussed shortly.)

The connective tissue layer is also richly supplied with blood vessels. They enter the bone at numerous locations and provide nutrients and oxygen and carry off cellular wastes such as carbon dioxide. The connective tissue layer is also supplied with many nerve fibers. The majority of the pain felt after a person bruises or fractures a bone results from stimulation of these pain fibers.

Inside many bones are cavities, known as **marrow cavities** (MARE-row). In fetuses, babies, and young children, the marrow cavities in most of the bones contain red marrow, so named because of its color. **Red marrow** produces red blood cells, white blood cells, and platelets to replace those lost each day. As an individual ages, most red marrow is slowly "retired" and becomes filled with fat, becoming **yellow marrow**. Yellow marrow begins to form during adolescence and, by adulthood, is present in all but a few bones. Red blood cell formation, however, continues in the bodies of the vertebrae, the hip bones, and a few others.

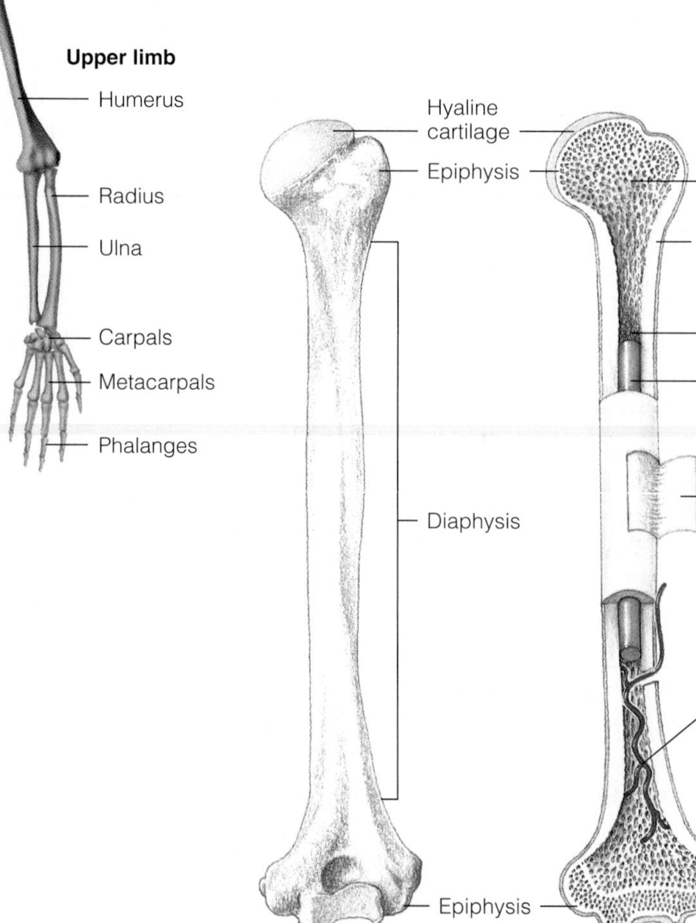

FIGURE 11-1 The Human Skeleton Over 200 bones of all shapes and sizes make up the human skeleton. The cartilage is shaded blue.

FIGURE 11-2 Anatomy of Long Bones *(a)* Drawing of the humerus in the upper arm. Notice the long shaft and dilated ends. *(b)* Longitudinal section of the humerus showing compact bone, spongy bone, and marrow.

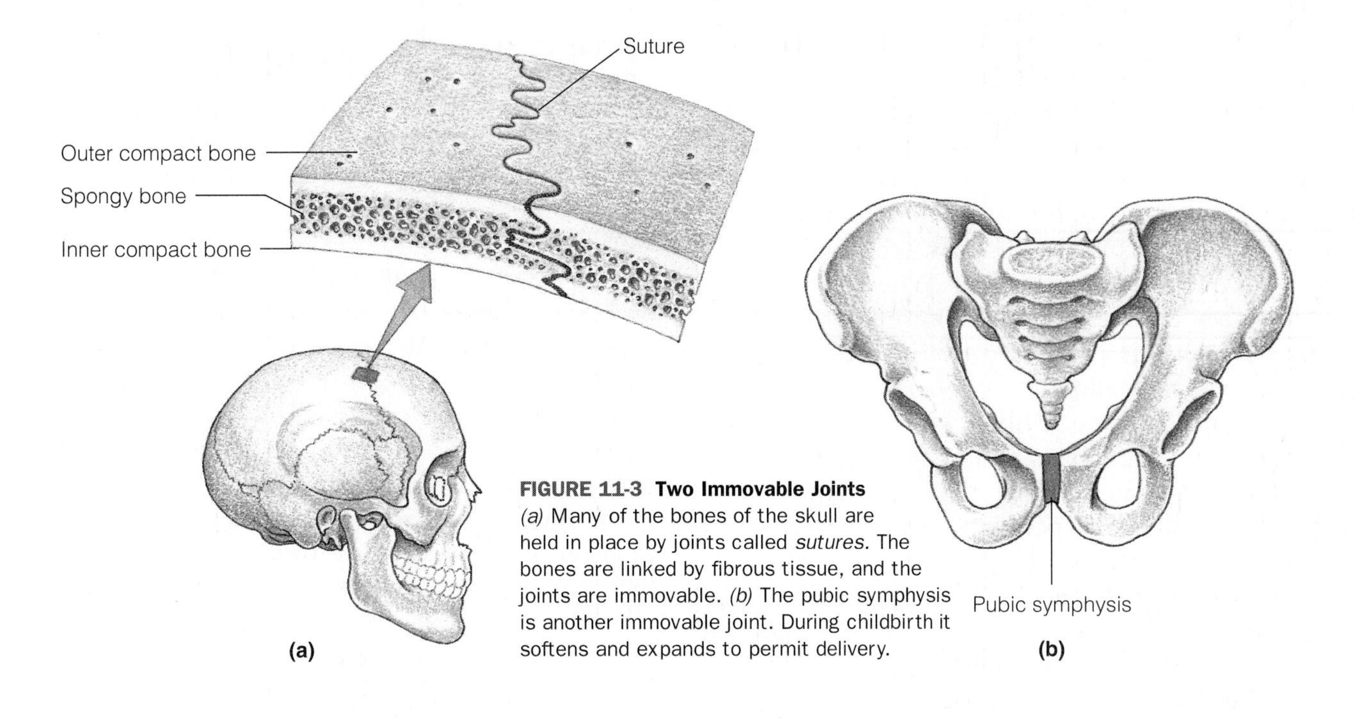

FIGURE 11-3 Two Immovable Joints *(a)* Many of the bones of the skull are held in place by joints called *sutures*. The bones are linked by fibrous tissue, and the joints are immovable. *(b)* The pubic symphysis is another immovable joint. During childbirth it softens and expands to permit delivery.

The Joints

Movements we make are made possible by the action of muscles on bones. But they could not occur without our joints. **Joints** are the structures that connect the bones of the skeleton. They can be classified by the degree of movement they permit from immovable to highly movable.

The bones of the skull are shown in **FIGURE 11-3A**. As illustrated, these bones interlock to form immovable joints. Fibrous connective tissue spans the space between the interlocking bones, holding them together.

Another immovable joint holds the pubic bones together—pubic symphysis (PEW-bick SIM-fa-siss), (**FIGURE 11-3B**). A special type of cartilage locks the bones together. However, in women near the end of pregnancy, hormones loosen the cartilage, allowing the pubic bones to separate. This allows the pelvic outlet to widen enough to permit a baby to pass through.

The bodies of the vertebrae are united by slightly movable joints (**FIGURE 11-4**). Each vertebra is separated from its nearest neighbors by an **intervertebral disc**. The inner portion of the disc acts as a cushion, softening the impact of walking and running. The outer, fibrous portion holds the disc in place and joins one vertebra to its nearest neighbor. The joints between the vertebrae, although only slightly movable, permit a fair amount of flexibility. If they did not, we would be unable to bend over to tie our shoes.

The most common type of joint is the freely movable, or **synovial joint** (sin-OH-vee-al). Although synovial joints differ considerably in architecture, they share several features. The first is the hyaline cartilage located on the joint surfaces of the bones (**FIGURE 11-5A**). This thin cap of slippery cartilage reduces friction and facilitates movement. The second commonality is the joint capsule, connective tissue that joins one bone to another (**FIGURE 11-5A**). In many joints, parallel bundles of dense connective tissue fibers in the outer layer of the capsule form **ligaments**. A ligament is a structure that joins bones to bones together in joints. They provide additional support to the joint.

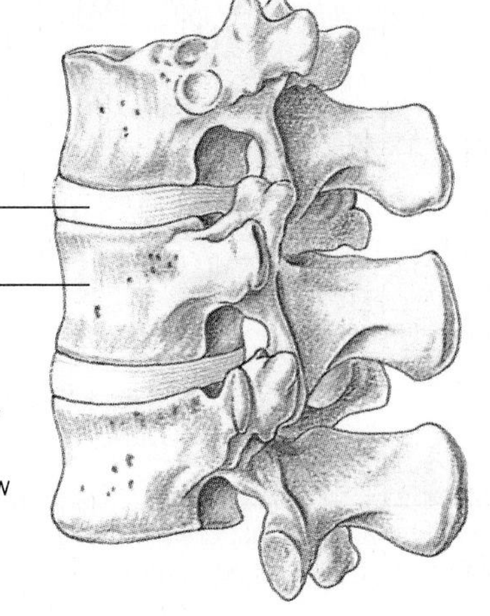

FIGURE 11-4 A Slightly Movable Joint The intervertebral discs allow for some movement, giving the vertebral column flexibility.

As a rule, ligaments are fairly inflexible. However, some individuals have remarkably flexible ligaments. They also have flexible **tendons**. (A tendon is the connective tissue that joins muscles to bones.) Because of this, some people exhibit extraordinary flexibility. For example, they may be able to extend their thumbs well beyond the 90 degrees possible for most of us. These people are said to be "double-jointed," a misnomer because the flexibility is permitted by the tendons and ligaments, not the joint.

The joint capsule is filled with a fairly thick, slippery fluid. It acts as a lubricant. Injuries to a joint, can result in a dramatic increase in this fluid, causing swelling and pain in joints.

The joint capsule and ligaments support joints. Additional support is provided by surrounding tendons and muscles. In the shoulder, for example, muscles help hold the head of the arm bone (humerus) in the socket. Because of this, individuals who are in poor physical shape are much more likely to suffer a dislocation.

A **dislocation** is an injury in which a bone is displaced from its proper position in a joint due to a fall or an unusual body movement. In some cases, bones slip out of place, then back in without assistance. In others, the bone must be put back in place by a trained health care worker.

Joint Injuries and Diseases

Joint injuries are common among athletes and physically active people. A hard blow to the knee of a football player, for instance, can tear the ligaments inside the knee joint or in the joint capsule.

Torn ligaments, tendons, and cartilage in joints heal very slowly because they have few blood vessels to bring nutrients needed for repair. For years, joint repair required major surgery and several months for recovery. Today, however, new surgical techniques allow physicians to repair joints with much less trauma. Through small incisions in the skin over the joint, surgeons insert a device called an **arthroscope** (ARE-throw-scope). It allows them to view the damage and to insert instruments to repair it without opening the joint. People are back on their feet in a few days.

Fractures

If you're like most people, chances are that sometime in your life you'll break a bone. Bone fractures can vary considerably in their severity. Some only involve hairline cracks, which mend fairly quickly. Others involve considerably more damage—a complete break, for instance—that

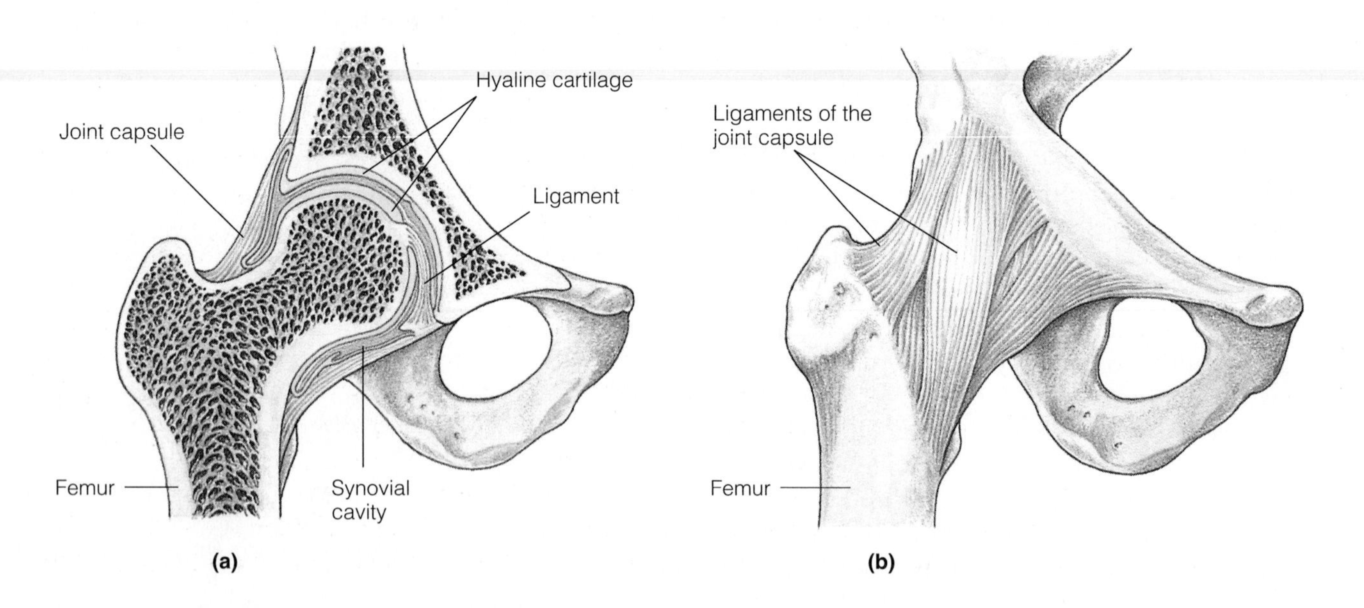

FIGURE 11-5 A Synovial Joint *(a)* A cross section through the hip joint (a ball-and-socket joint) showing the structures of the synovial joint. *(b)* Ligaments in the outer portion of the joint capsule help support the joint.

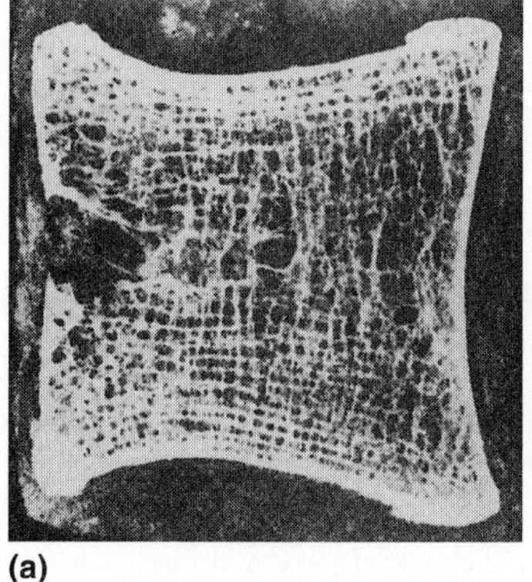
(a)

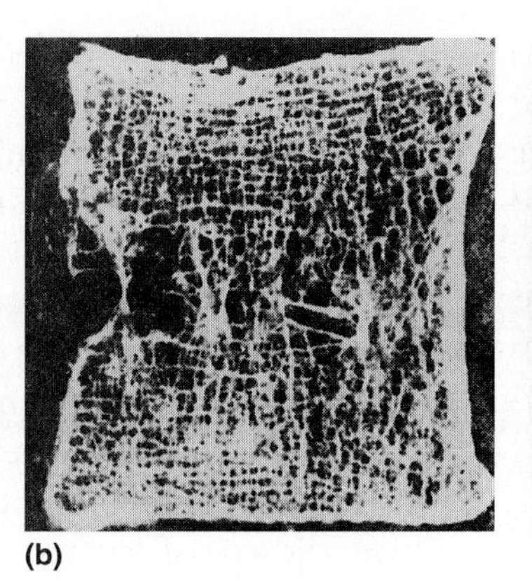
(b)

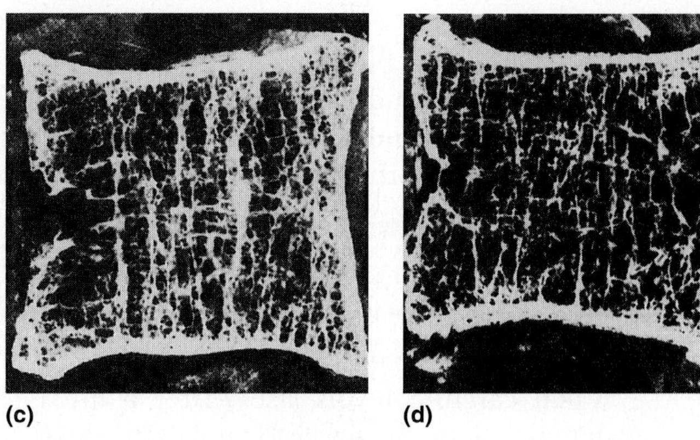
(c) (d)

FIGURE 11-6 Osteoporosis The loss of estrogen or prolonged immobilization weakens bone. In these situations, bone is dissolved and becomes brittle and easily breakable. (a) A section of the body of a lumbar vertebra from a 29-year-old woman. (b) Some thinning is evident in a vertebra of a 40-year-old woman. (c) Bone loss is severe in an 84-year-old woman. (d) Bone loss is most severe in a 92-year-old woman.

takes much longer to repair. In some instances, the broken ends must be realigned, even pinned together. A cast is then applied to hold the bones in position so they can heal.

Degenerative Arthritis

Over time, joints can wear out or degenerate. The result is **degenerative joint disease**, also known as **osteoarthritis**. This disease results from wear and tear on a joint. Over time, it causes the cartilage on the ends of bones to flake and crack. As the cartilage degenerates, the bones come in contact and grind against each other. This causes considerable swelling, pain, and discomfort.

Osteoarthritis occurs most often in the weight-bearing joints—the knee, hip, and spine—joints subject to the most wear over time. Osteoarthritis is extremely common in people over 40. Fortunately, many people do not even notice it, and the disease rarely becomes a serious medical problem.

Wear and tear on joints is worsened by obesity. The extra pressure wears the cartilage away more quickly. For such individuals, weight control can reduce the rate of degeneration and ease the pain. Painkillers such as aspirin and other anti-inflammatory drugs are used to treat the pain and swelling.

Rheumatoid Arthritis

Another common disorder of the synovial joint is rheumatoid arthritis (RUE-mah-toid). **Rheumatoid arthritis** is the most painful and crippling form of arthritis. It is caused by an inflammation of the inside lining of the joint capsule that often spreads to the cartilage. If the condition persists, rheumatoid arthritis can wear through the cartilage and cause degeneration of the underlying bone. Thickening of the lining of the joint capsule and degeneration of the bone often disfigure the joints. Patients lose movement in the joint and often suffer considerable pain. Joints can be completely immobile. Rheumatoid arthritis generally occurs in the joints of the wrist, fingers, and feet.

Rheumatoid arthritis may result from an autoimmune reaction. Rheumatoid arthritis occurs in people of all ages, but most commonly appears in individuals between the ages of 20 and 40. It is usually a permanent condition, although the degree of severity varies widely. Patients suffering from it can be treated with physical therapy, painkillers, anti-inflammatory drugs, and surgery.

Diseased joints can also be replaced by artificial joints that restore mobility and reduce pain. Plastic joints are used to replace the finger joints, greatly improving the appearance of the hands and restoring use of the fingers. Severely damaged knee and hip joints are replaced with special steel or Teflon artificial joints.

Regulating Calcium Levels

Bone also helps to regulate calcium levels in the blood. When blood levels are high, calcium is stored in the bone. When blood levels fall, calcium is released back into the blood.

The uptake and release are controlled by two hormones. When blood calcium levels fall, for instance, small glands embedded on the back side of the thyroid gland, known as the **parathyroid glands**, release a hormone known as **parathormone** (PAIR-ah-THOR-moan) into the bloodstream. This hormone stimulates the bone-destroying cells called *osteoclasts*. They digest bone in their vicinity, releasing calcium into the bloodstream. This restores blood calcium levels.

When calcium levels rise—after a meal for example—cells in the thyroid gland itself release the hormone **calcitonin** (CAL-seh-TONE-in). This hormone inhibits the bone-destroying cells. It also stimulates bone-forming cells, causing them to deposit new bone. The result is that blood calcium levels fall, returning to normal.

Osteoporosis

One of the most common problems adults face is **osteoporosis** (OSS-tea-oh-puh-ROE-siss; "porous bone"). This disease is characterized by a progressive loss of bone calcium and weakening of the bone (**FIGURE 11-6**). In some individuals, calcium loss is so severe that bones become extremely brittle and break as a result of normal activities such as getting out of bed in the morning.

Osteoporosis occurs most often in women after menopause. **Menopause** (MEN-oh-pause) usually occurs between 45 and 55 years of age. It results from the loss of estrogen production by a woman's ovaries. Although estrogen is primarily a reproductive hormone, it also helps to maintain bone. Osteoporosis also occurs in people who are immobilized for long periods, for example in a hospital bed. Osteoporosis is not inevitable and can be prevented by exercise, calcium supplements, estrogen supplements, and other measures.

Bone is an important tissue. Besides providing support and protection, our bones help ensure homeostasis.

The Muscular System

The muscles of our body perform many functions. As shown in FIGURE 12-1, many muscles cross joints. When these muscles contract, they produce movement. But not all skeletal muscles make bones move. Some muscles simply steady joints. These muscles help to maintain our posture, permitting us to stand or sit upright despite the constant pull of gravity. Muscles beneath the skin of the face allow us to wrinkle our skin, open and close our eyes, and move our lips. But that's not all. Muscles also produce enormous amounts of heat as a by-product of metabolism. Working muscles produce even more heat—so much, in fact, that you can cross-country ski in freezing weather wearing only a light sweater.

In this module, we'll examine the structure and function of muscles.

Understanding Muscle Contraction

Skeletal muscles consist of bundles of long, unbranched cells, called *muscle fibers*. Muscle fibers are formed during embryonic development by the fusion of many cells. Viewed with the light microscope, skeletal muscle fibers appear banded (FIGURE 12-2).

Like nerve cells, the muscle fiber is an excitable cell. Muscle fibers are stimulated by nerve impulses arriving at the terminal boutons. Nerve impulses stimulate the release of a neurotransmitter substance that stimulates an impulse in the muscle fiber itself. This new impulse travels along the membrane of the muscle fiber causing the cell to contract.

Contraction is made possible by contractile proteins in muscle fibers. When stimulated, the contractile proteins cause the cells to shorten, making the muscle contract and perform work.

To understand how these proteins work, let's imagine that you could isolate a single skeletal muscle fiber (cell) free from a skeletal muscle (FIGURE 12-3A). Inside each muscle fiber are numerous smaller fibers, called **myofibrils** (MY-oh-FIE-brills) (FIGURE 12-3B on page 92). Myofibrils contain contractile proteins. You will also notice that each myofibril is banded. The banded appearance is due to the arrangement of the contractile proteins.

FIGURE 12-1 The Skeletal Muscles

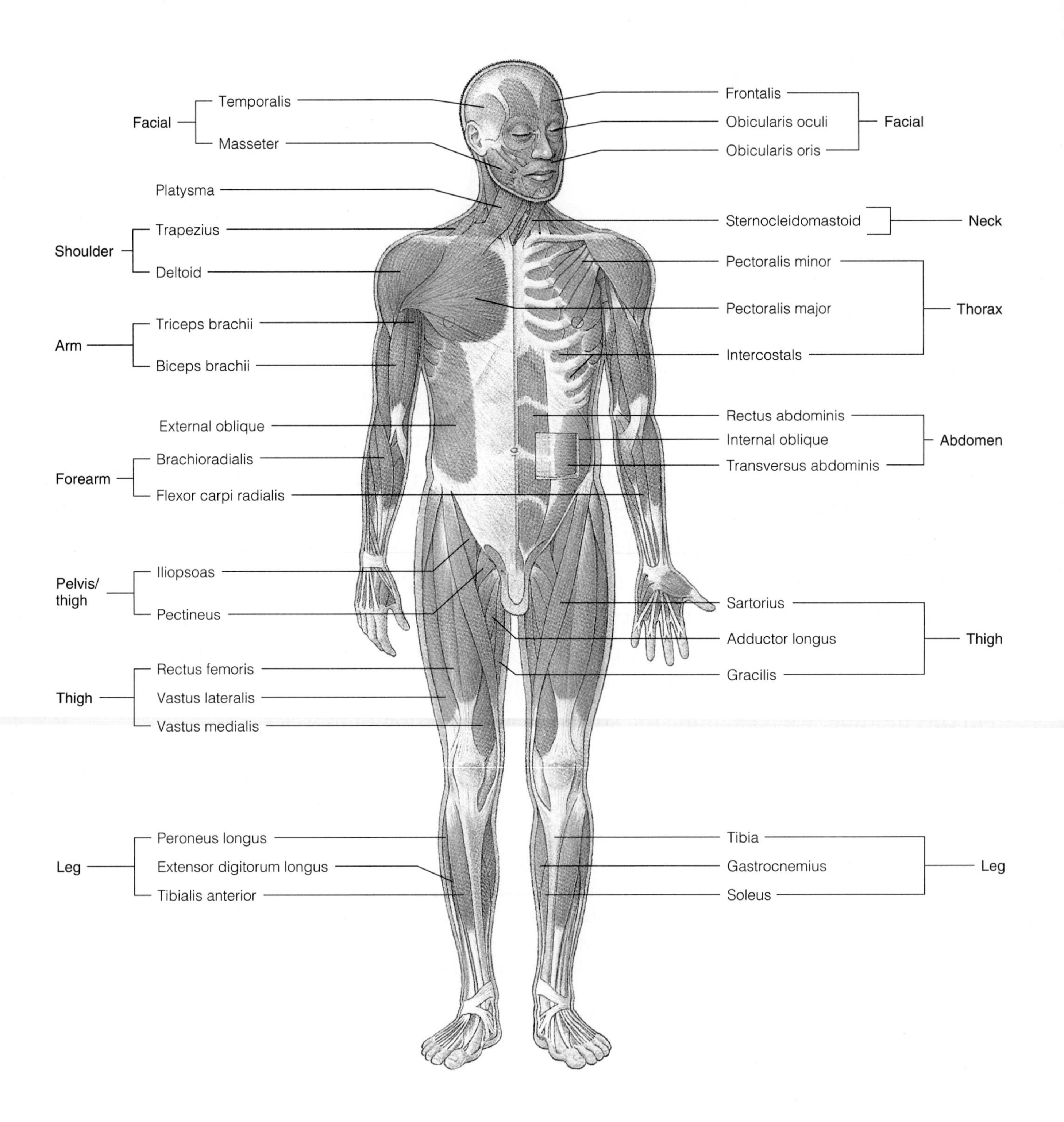

FIGURE 12-3D shows that each myofibril contains two types of contractile protein, the thick and thin filaments. The thick filaments consist of the protein **myosin** (MY-oh-sin). The thin filaments are composed primarily of the protein **actin** (ACK-tin).

Muscle contraction is achieved by the sliding of actin filaments over the myosin filaments. Actin filaments are actually pulled by the thick filaments (myosin molecules). How is this accomplished?

As **FIGURE 12-4** shows, each thick filament actually consists of numerous golf club-shaped myosin molecules arranged with their "club ends," or heads, projecting toward the actin filaments. During muscle contraction, the heads of the myosin molecules attach to actin filaments. They then tug the actin filaments inward, causing the myofibril to shorten and the muscle to contract.

The attachment of the myosin molecules to the actin molecules is stimulated by the release of calcium inside muscle cells. That's one reason why calcium is so important. Calcium is stored in the smooth endoplasmic reticulum of muscle cells. The release of calcium, in turn, is stimulated by electrical impulses from the nerves.

Muscle contraction also requires energy. It comes from ATP. ATP is produced by the breakdown of glucose in cells.

Fast- and Slow-Twitch

Have you ever wondered why some athletes excel at events that require rapid explosion of energy, such as sprints, while others excel at events like marathons, which require more endurance?

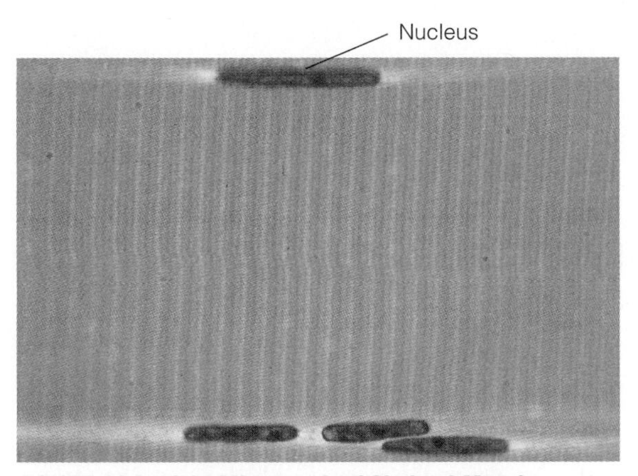

FIGURE 12-2 Light Micrograph of Skeletal Muscle
Notice the banding pattern on these muscle fibers.

Research on muscle physiology shows that human skeletal muscles contain two types muscle fibers: fast-twitch and slow-twitch fibers. **Slow-twitch muscle fibers** contract relatively slowly but have incredible endurance. They're found in greater amounts in the muscles of endurance athletes such as long-distance runners. These permit such athletes to perform for long periods without tiring.

Fast-twitch muscle fibers contract swiftly. The muscles of sprinters and other athletes whose performance depends on quick bursts of activity contain a high proportion of fast-twitch fibers.

Skeletal muscles generally contain a mixture of slow- and fast-twitch fibers, giving each muscle a wide range of performance abilities. However, a muscle that performs one type of function more often than another tends to have a higher number of fibers corresponding to the type of activity it performs. The muscles of the back, for example, contain a larger number of slow-twitch fibers. These muscles operate throughout the waking hours to maintain posture. They do not need to contract quickly, but they must be resistant to fatigue. In contrast, the muscles of the arm are used for many quick actions. Fast-twitch fibers are more common in the muscles of the arm.

The Benefits of Exercise

When muscles are made to work hard, they respond by becoming larger and stronger. The increase in size and strength results from an increase in the amount of contractile protein inside muscle cells. Unfortunately, muscle protein is quickly lost if exercise ceases. In fact, about half of the muscle you gain in a weight-lifting program is broken down two weeks after you stop exercising. If you don't keep up with your exercise program, your newly developed muscles won't last long.

High-intensity exercise, such as weight lifting, builds muscle. It takes surprisingly little exercise to have an effect. Working out every other day for only a few minutes will result in noticeable changes in muscle mass. According to one source, 18 contractions of a muscle (in three sets of six contractions each) are enough to increase muscle mass—if they force your muscles to exert over 75% of their maximum capacity.

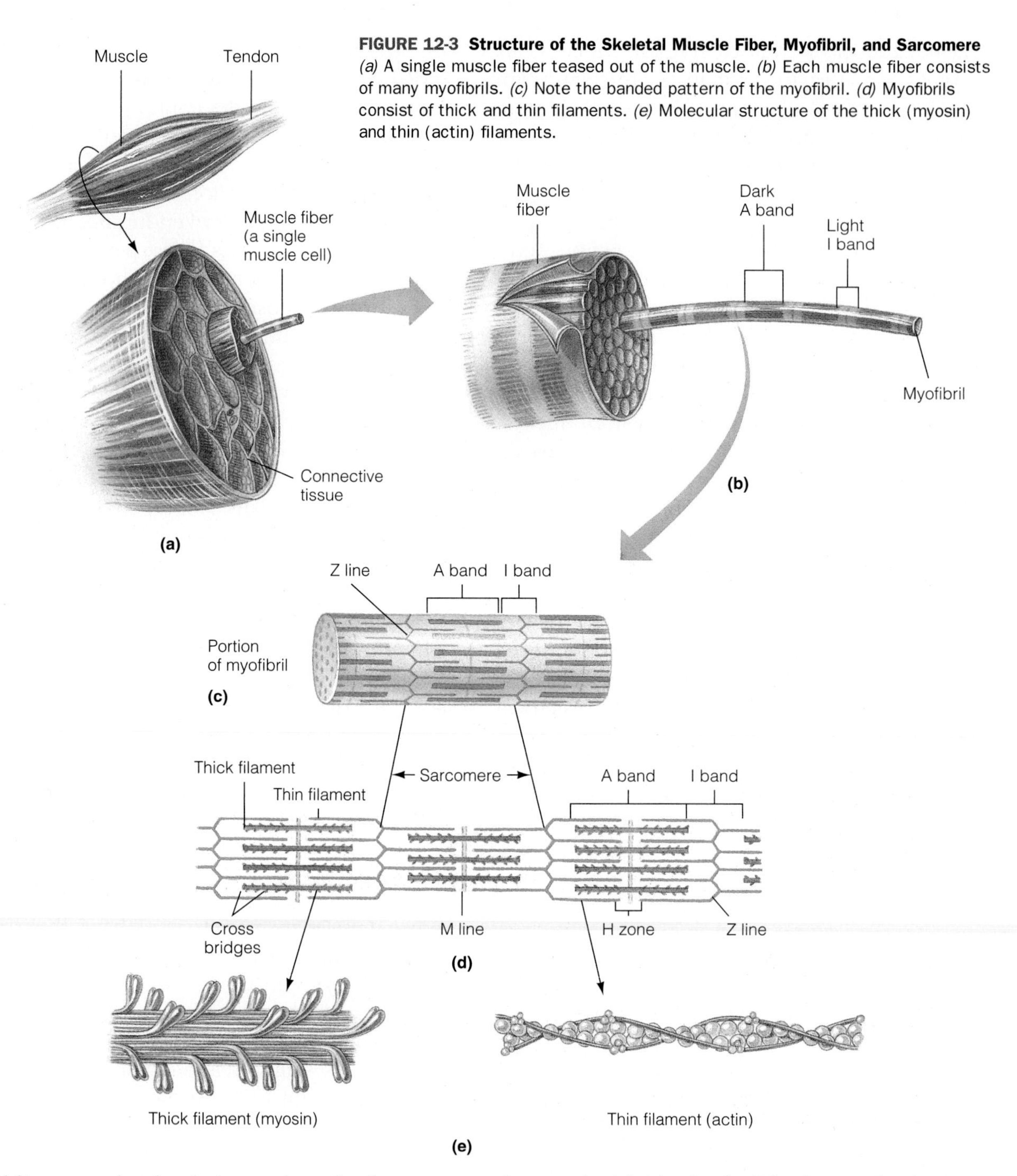

FIGURE 12-3 Structure of the Skeletal Muscle Fiber, Myofibril, and Sarcomere *(a)* A single muscle fiber teased out of the muscle. *(b)* Each muscle fiber consists of many myofibrils. *(c)* Note the banded pattern of the myofibril. *(d)* Myofibrils consist of thick and thin filaments. *(e)* Molecular structure of the thick (myosin) and thin (actin) filaments.

Building muscle also helps to lose fat for two reasons. First, exercise burns fat. Second, because muscle has a fairly high energy demand, even at rest, it will burn up fat when you're not exercising.

Low-intensity exercise, such as aerobics and swimming, tends to burn calories as well. However, it does not build muscle bulk. Aerobic exercise therefore helps increase endurance—that is, it increases one's ability to sustain muscular effort. Stamina or endurance results from numerous changes inside the body. The heart, for instance, grows stronger and enlarges when you work it harder. A well-exercised heart beats more slowly but pumps more blood with each beat. Thus, the heart works more efficiently, delivering more oxygen to skeletal muscles.

Increased endurance also results from improvements in the respiratory system. For example, exercise increases the strength of the muscles involved in breathing. When they are

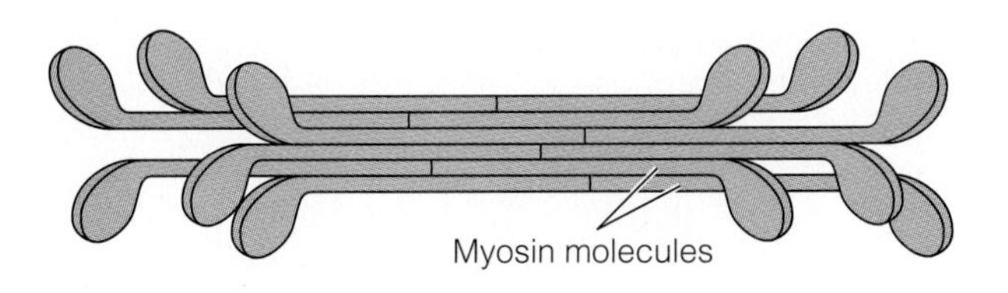

FIGURE 12-4 Structure of Myosin Filaments Myosin molecules join to form a myosin filament. Note the presence and orientation of the heads of the myosin molecules.

stronger, they can operate longer without tiring. Also, breathing during exercise becomes more efficient.

Increased endurance also results from an increase in the amount of blood resulting from exercise. An increase in blood volume results in an increase in the number of RBCs, which increases the amount of oxygen available to cells. This improvement, combined with others, allows an individual to work out longer without growing tired.

When you set out on an exercise program, it is important to establish your goals first. If you are interested in increasing your endurance, you should pursue an exercise regime that works the heart and muscles at a lower intensity over longer periods such as riding a bicycle or jogging. If you are after bulk, high-intensity exercise programs such as weight lifting will do.

Many health clubs offer a variety of machines to help you build muscles or simply tone them. Most of the exercise machines work one particular muscle group—for example, the muscles of the upper arm or the muscles of the chest. In most gyms, a half-dozen or more machines are usually placed in a line so you can go down the line, working one set of muscles after another until you have exercised your entire body.

Exercise machines are popular because they are safer than free weights (barbells and dumbbells). Progressive resistance machines eliminate the chances of your dropping a weight on your toes-or someone else's! Moreover, it is almost impossible to strain your back if you make a mistake using one. In contrast, lifting free weights requires care and training as well as brawn.

These machines also reduce the amount of time a person needs to exercise by about half because they've been designed to require work when a joint is both flexed and extended. The biceps machine, for example, requires you to pull the weights up, then return them slowly. Your muscles are being forced to work in both directions.

Staying in shape not only makes you look and feel better, it will help you live a longer life, by burning fat and keeping your heart and arteries from clogging with cholesterol. Good exercise programs will also reduce stress and help you sleep better. Twenty minutes three times a week is a good goal to set. As you get stronger you can devote more time to exercise. You won't be sorry.

Muscles of the body serve us well, but the benefits are not just the obvious motion they permit. Muscles generate heat and support joints, too, making them vital to our health and well being.

The Endocrine System

You've learned in previous modules that the nervous system controls many body functions. But there's another system that plays a key role in controlling body functions—the *endocrine system*.

What is the Endocrine System?

The endocrine system consists of numerous small glands scattered throughout the body (**FIGURE 13-1**). These glands produce and secrete chemical substances called *hormones* (HOAR-moans). A **hormone** is a chemical produced and released by cells or groups of cells that form endocrine (ductless) glands. Hormones are transported in the bloodstream to distant sites where they exert some effect. The cells affected by a hormone are called its *target cells*.

The blood carries dozens of hormones at any one time. These hormones function in five areas: (1) homeostasis; (2) growth and development; (3) reproduction; (4) energy production, storage, and use; and (5) behavior.

Target Cells, Receptors, and Cellular Responses

Despite the fact that they're exposed to many different hormonal signals, target cells respond only to specific signals. The reason for this is that target cells contain protein receptors that bind to specific hormones. Each cell contains receptors for the hormones it is genetically programmed to respond to. In some target cells, the hormone receptors are in the cell membrane; in others, they're located in the cytoplasm.

Hormones fall into two broad categories. The first are the tropic hormones (TROW-pick; "to nourish"). **Tropic hormones** stimulate the production and secretion of hormones by other endocrine glands. An example is thyroid-stimulating hormone (TSH). Produced by the pituitary gland, TSH travels in the blood to the thyroid gland in the neck on either side of the voice box. Here TSH stimulates the release of another hormone, thyroxine (thigh-ROX-in).

Thyroxine, in turn, circulates in the blood and stimulates metabolism in many types of body cells. Thyroxine is a **nontropic hormone.** Nontropic hormones stimulate cellular growth, metabolism, or other functions.

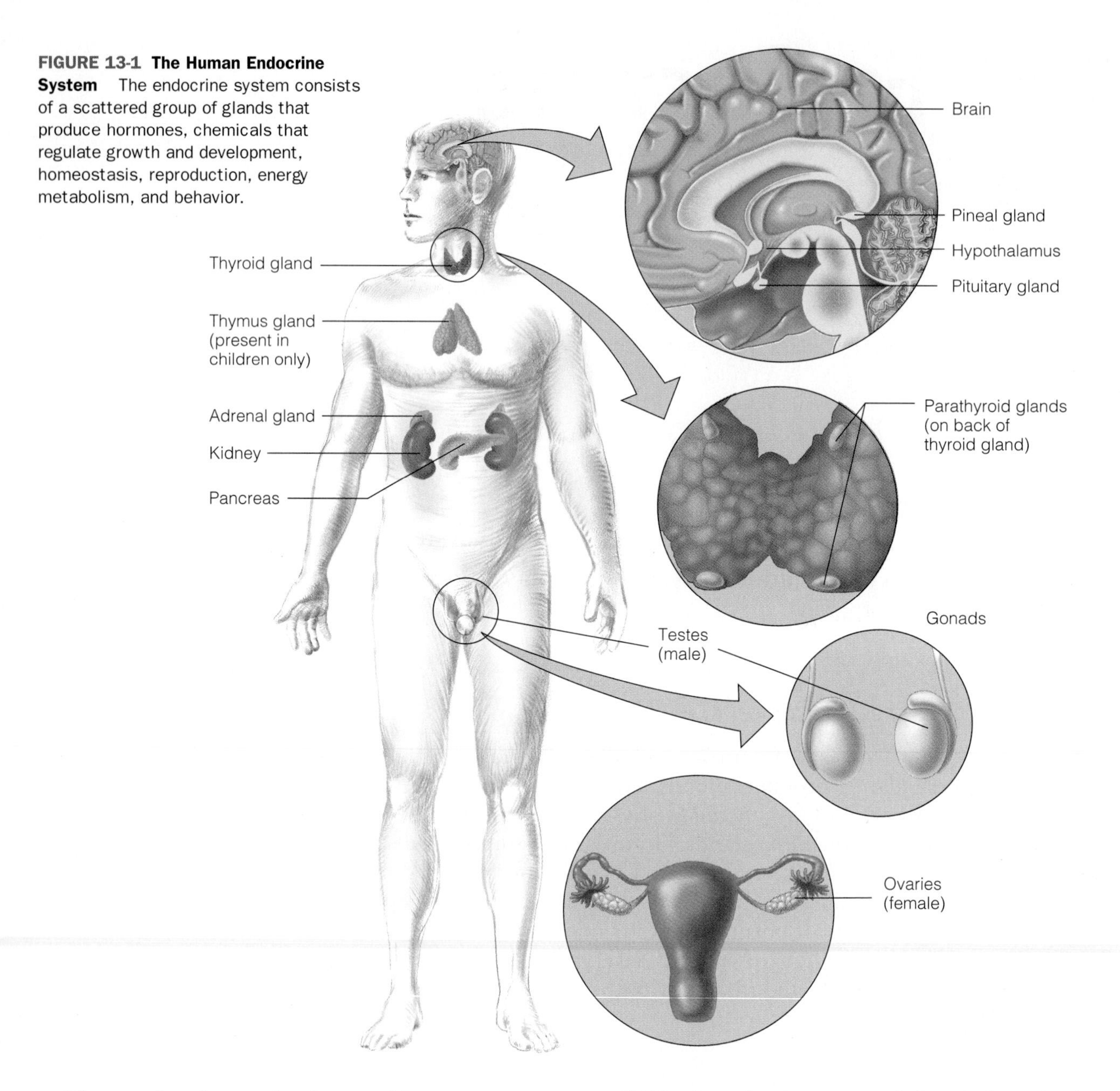

FIGURE 13-1 The Human Endocrine System The endocrine system consists of a scattered group of glands that produce hormones, chemicals that regulate growth and development, homeostasis, reproduction, energy metabolism, and behavior.

The production and release of hormones are controlled by negative feedback, described in Module 1. Negative feedback maintains levels necessary for proper functioning of the body. You'll see many examples of this phenomenon in this module.

The fact that hormones are controlled by negative feedback loops does not mean that hormone concentrations in the blood are constant 24/7. In fact, virtually all hormones undergo daily changes in their release, causing many body functions to fluctuate. These natural fluctuations in body function are called *biological cycles*, or *biorhythms*, described in Module 1.

With these basics in mind, let's take a look at the main endocrine glands and the hormones they produce.

The Pituitary and Hypothalamus

Attached to the underside of the brain by a thin stalk is the **pituitary gland** (peh-TWO-eh-TARE-ee) (**FIGURE 13-2**). It's about the size of a pea.

The pituitary gland is divided into two parts: the **anterior pituitary** and the **posterior pituitary**. Together, they secrete a large number of hormones, affecting a great many of the body's functions. Let's begin with the anterior pituitary.

The anterior pituitary produces seven hormones, six of which are discussed in this module. The production and release of these hormones are controlled by a region of the brain just above the pituitary, known as the *hypothalamus*.

The hypothalamus contains receptors that monitor blood levels of hormones, nutrients, and ions. When activated, the receptors stimulate specialized nerve cells within the hypothalamus. These cells are a special brand of neurons called **neurosecretory neurons**. They get this name from that fact that they synthesize and secrete (release) hormones. The hormones produced by these cells control the production and release of hormones by its near neighbor, the anterior pituitary (**FIGURE 13-2**).

The hypothalamus produces two types of hormones—those that stimulate the release of pituitary hormones, called *releasing hormones*, designated RH, and those that inhibit the release of hormones from the anterior pituitary. They're called *inhibiting hormones*, designated IH.

The releasing and inhibiting hormones are released into the bloodstream in the hypothalamus. From here they travel in a special network of blood vessels to the pituitary (**FIGURE 13-2**). The hormones bind to their target cells, causing them to release their hormones.

Growth Hormone. The anterior pituitary controls growth through the release of an appropriately named hormone, growth hormone. **Growth hormone** (GH) is a protein hormone. It stimulates growth by promoting an increase in the size and the number of cells (through cell division). Although growth hormone affects virtually all body cells, it acts primarily on bone and muscle.

As a rule, the more growth hormone that is produced when one is growing, the taller and heftier he or she will be. As **FIGURE 13-3** shows, the highest blood levels are present during sleep and during strenuous exercise. It is no wonder that sleep is so important to a growing child. Growth hormone secretion decreases gradually as we age.

The secretion of growth hormone is controlled by a releasing hormone (GH-RH) produced by the hypothalamus. Growth hormone participates in a classic negative feedback loop. When levels are low, GH-RH is released from the hypothalamus, causing the anterior pituitary to produce and release more GH. Growth hormone release is also stimulated directly through the nervous system. Studies show that stress and exercise both stimulate the hypothalamus to release GH-RH.

Deficiencies in growth hormone can result in dramatic changes in body shape and size. If the deficiency occurs during the growth phase, a child may be stunted. This condition is known as

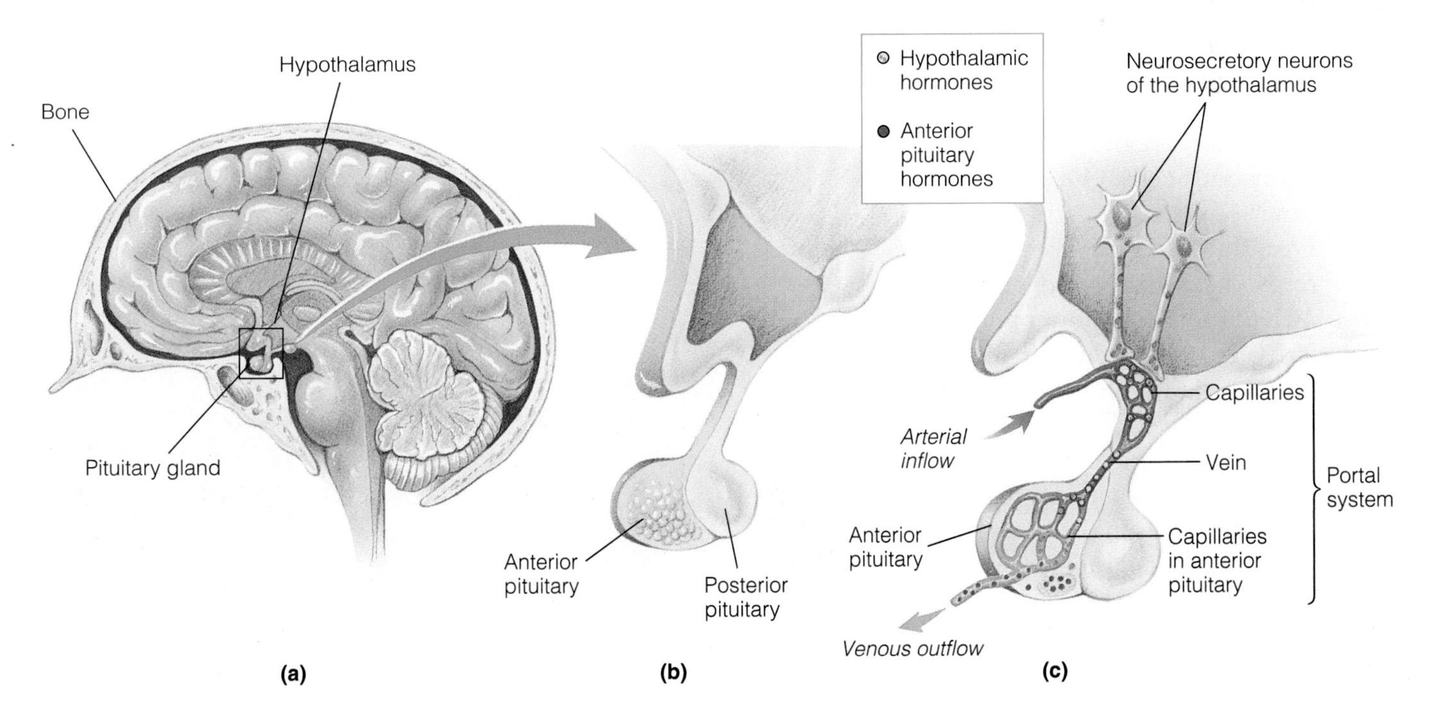

FIGURE 13-2 The Pituitary Gland *(a)* A cross section of the brain showing the location of the pituitary and hypothalamus. *(b)* The structure of the pituitary gland. *(c)* Releasing and inhibiting hormones travel from the hypothalamus to the anterior pituitary, where they affect hormone secretion.

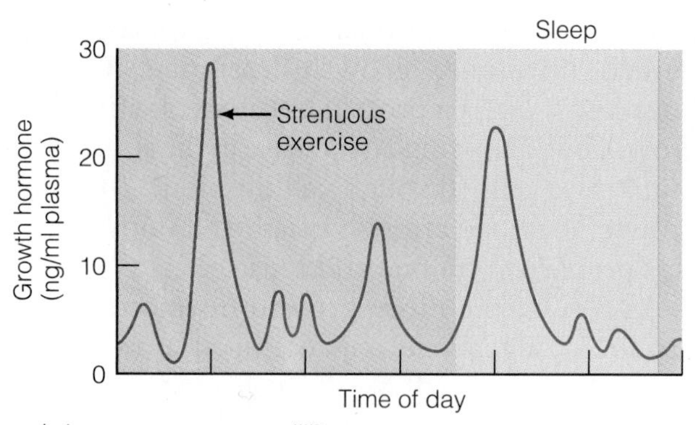

FIGURE 13-3 Growth Hormone Secretion in an Adult
Growth hormone stimulates muscle and bone growth and is released during exercise and at night.

dwarfism (**FIGURE 13-4A**). If the excess occurs during the growth phase, oversecretion results in **giantism** (GIE-an-tizm) (**FIGURE 13-4B**).

Thyroid-Stimulating Hormone. **Thyroid-Stimulating Hormone** (**TSH**) is a protein hormone produced by the anterior pituitary. It is controlled by the hypothalamus. TSH release is regulated by the level of thyroid hormone, thyroxine, in the blood in a classical negative feedback loop. Receptors in the hypothalamus detect the level of thyroxine. When levels are low, these receptors signal the hypothalamus to release TSH-RH. TSH causes the release of thyroid hormones. When the level of thyroxine increases, TSH-RH secretion declines (**FIGURE 13-5**). TSH-RH secretion, however, is also stimulated by cold and stress.

Thyroid hormones circulate in the bloodstream and influence many body cells. One of their chief functions is to stimulate the breakdown of glucose by body cells to produce energy and heat.

ACTH. Another important hormone is **ACTH** or **adrenocorticotropic hormone** (ad-REE-no-core-tick-oh-TRO-pick). ACTH is produced by the anterior pituitary. Its target organ is a layer of hormone-producing cells in the outer regions of the adrenal gland, known as the *adrenal cortex*. These cells produce a group of steroid hormones known as the **glucocorticoids** (GLUE-co-CORE-teh-KOIDS). Their main function is to increase blood glucose levels, thus helping maintain homeostasis.

As in other anterior pituitary hormones, ACTH secretion is controlled by the hypothalamus. As shown in **FIGURE 13-6**, the hypothalamus controls ACTH production and release via ACTH-RH.

ACTH-RH secretion is also controlled by stress. When we're under stress, ACTH-RH secretion increases. This causes an increase in ACTH release that stimulates an increase in the release of glucocorticoids by the adrenal cortex. Glucocorticoids increase blood glucose levels, which ensures that we have the additional energy we need to operate body systems, especially muscles, when we are under stress.

The Gonadotropins. Reproduction in both males and females is primarily under the control of the anterior pituitary. It produces two hormones that affect the gonads, appropriately known as *gonadotropins* (go-NAD-oh-TROW-pins). These hormones are discussed in the next module.

Prolactin. Milk production in women is stimulated by the hormone **prolactin** (pro-LACK-tin). Prolactin secretion, shown in **FIGURE 13-7**, is

(a) (b)

FIGURE 13-4 Disorders of Growth Hormone Secretion
(a) Pituitary dwarves. *(b)* Pituitary giant.

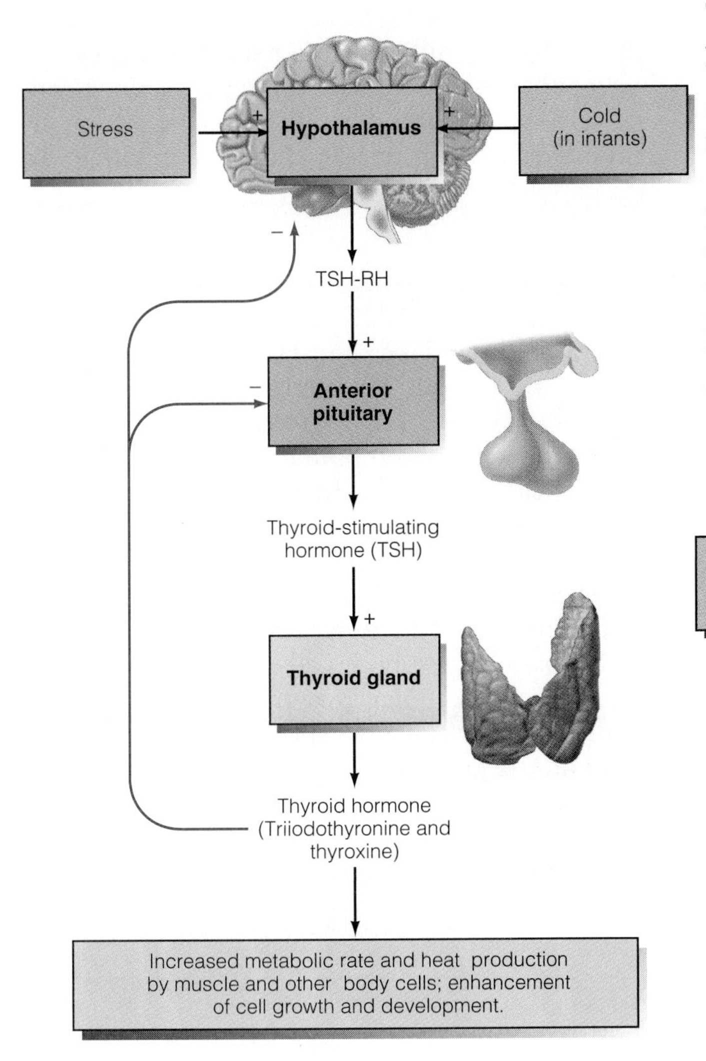

FIGURE 13-5 Negative Feedback Control of TSH Secretion Thyroxine levels are detected by receptors in the hypothalamus. When levels are low, the hypothalamus releases TSH-RH, which stimulates the pituitary to release TSH. Other factors such as stress and cold influence the release of TSH via the hypothalamus. (A + denotes stimulation; a – denotes inhibition.)

stimulated by suckling. During suckling, nerve impulses travel from the breast to the hypothalamus. Here, they stimulate the release of prolactin releasing hormones. It travels to the anterior pituitary in the bloodstream, where it stimulates the secretion of prolactin.

Prolactin production continues as long as suckling continues. As babies begin to eat solid food, however, reduced suckling shuts down prolactin secretion and the breasts cut back on milk production.

The Posterior Pituitary

The **posterior pituitary** produces two hormones: **oxytocin** (OX-ee-TOE-sin) and **antidiuretic hormone** (an-tie-DIE-yur-ET-ick) (ADH).

As shown in **FIGURE 13-8**, ADH and oxytocin are produced in the cell bodies of the neurosecretory cells located in the hypothalamus. They then travel down the axons of these nerves into the posterior pituitary. The hormones are stored in the ends of the nerves until being released into the surrounding capillaries.

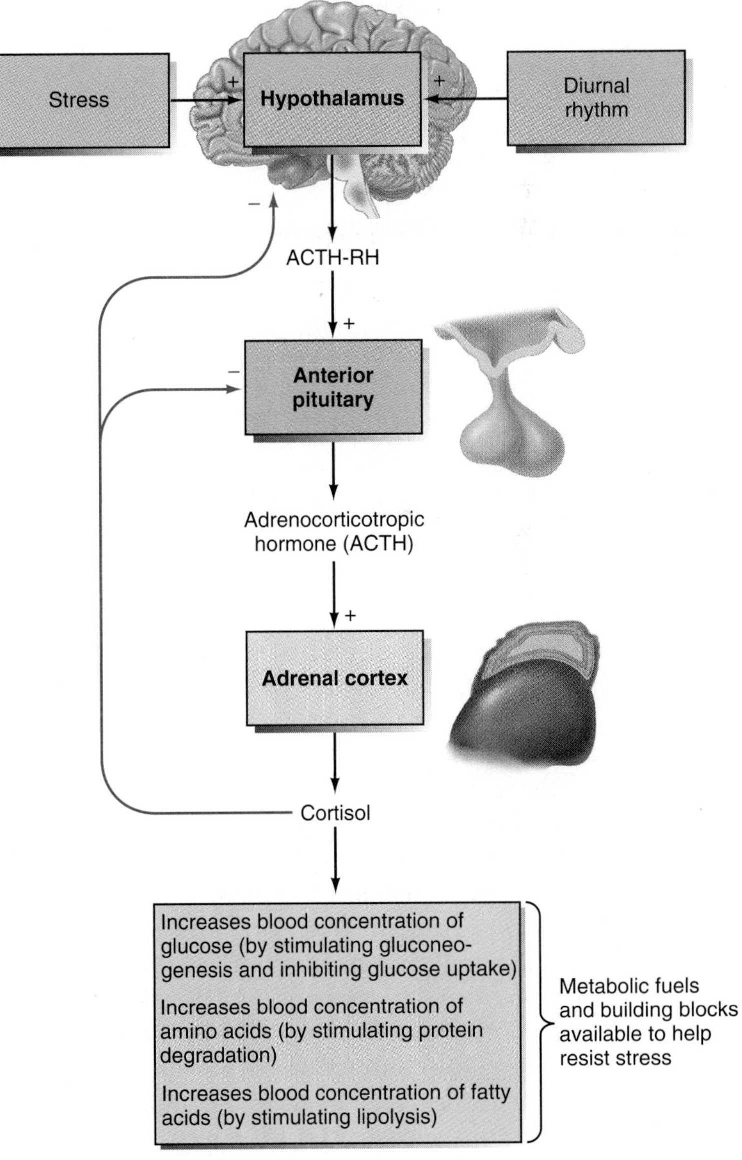

FIGURE 13-6 Feedback Control of ACTH Cortisol regulates hypothalamic and pituitary activity, but stress and the biological clock also influence the release of ACTH-RH.

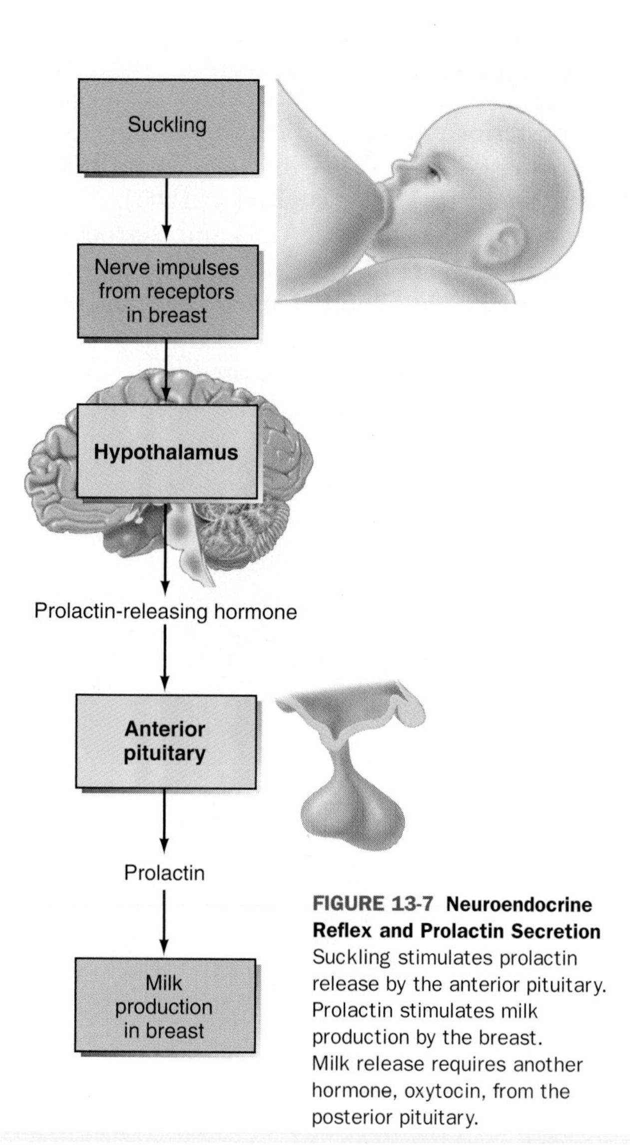

FIGURE 13-7 Neuroendocrine Reflex and Prolactin Secretion Suckling stimulates prolactin release by the anterior pituitary. Prolactin stimulates milk production by the breast. Milk release requires another hormone, oxytocin, from the posterior pituitary.

Antidiuretic Hormone. As explained in Module 8, ADH regulates water balance in humans by increasing water absorption in the nephrons of the kidneys (**FIGURE 13-9**). As a result, water reenters the bloodstream, increasing blood volume and maintaining the normal concentration of chemicals in the blood—and blood pressure. Control of ADH release is explained in Module 8.

Oxytocin. Prolactin from the anterior pituitary stimulates milk production in the breasts. But it's oxytocin from the posterior pituitary that actually "pumps" the milk from the glands. It does this by stimulating the contraction of the smooth-muscle-like cells around the glands in the breast.

Oxytocin release is stimulated by suckling. Sensory fibers in the breast conduct nerve impulses to the hypothalamus, triggering the release of oxytocin in the posterior pituitary. The hormone travels in the blood to the breast, where it stimulates the ejection of milk soon after suckling begins.

Oxytocin is also released during birth. Oxytocin travels in the blood to the uterus, where it stimulates smooth muscle contraction, aiding in the expulsion of the baby.

The Thyroid Gland

The **thyroid gland** is a U- or H-shaped gland in the neck (**FIGURE 13-10**). The thyroid gland produces three hormones: (1) thyroxine or T4, (2) a chemically similar compound, called T_3 for short, and (3) calcitonin. The first two are involved in controlling metabolism and heat production, as explained earlier. Calcitonin helps to regulate blood levels of calcium.

Thyroxine and T_3 accelerate the rate of glucose breakdown in most body cells. Thyroid hormones also stimulate cellular growth and development. Bones and muscles are especially dependent on them during growth.

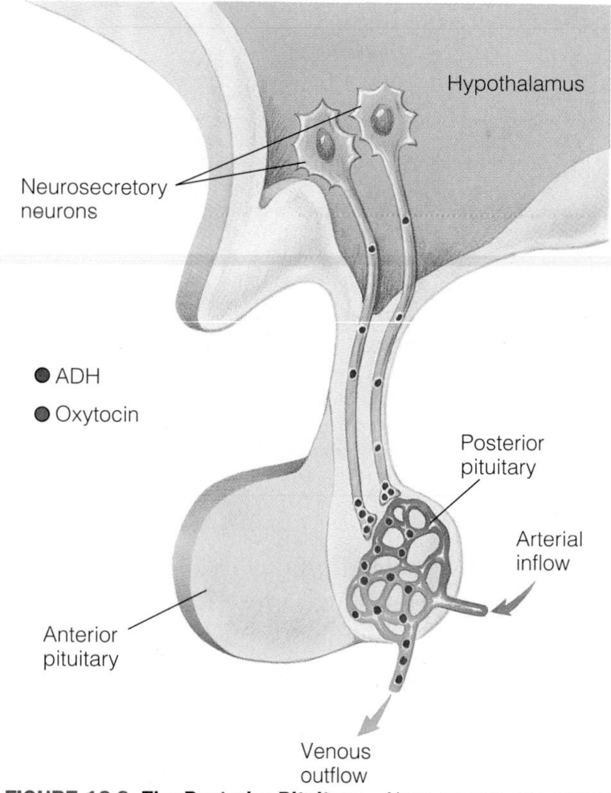

FIGURE 13-8 The Posterior Pituitary Neurosecretory neurons that produce oxytocin and ADH originate in the hypothalamus and terminate in the posterior pituitary. Hormones are produced in the cell bodies of the neurons and are stored and released into the bloodstream in the posterior pituitary.

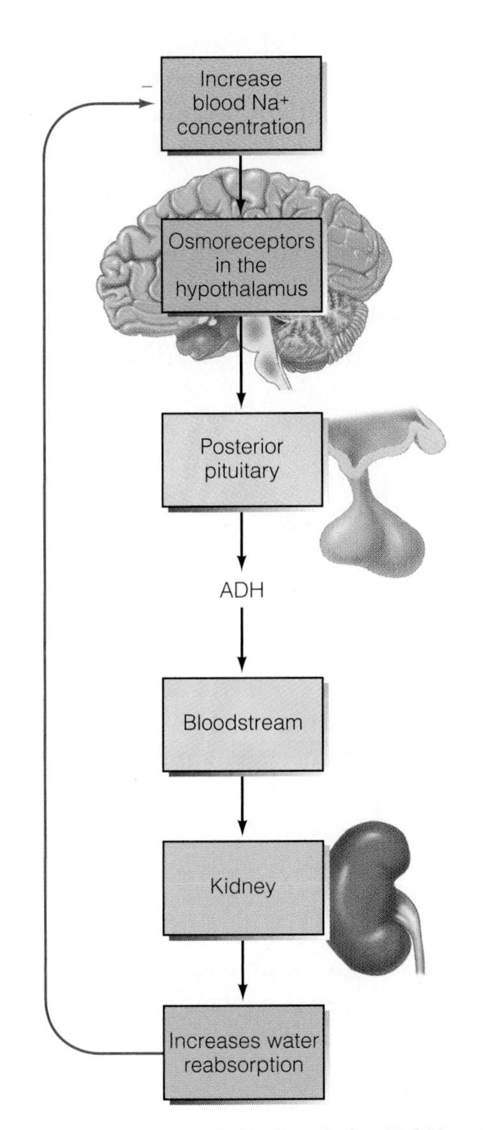

FIGURE 13-9 Role of ADH in Regulating Fluid Levels ADH secretion is stimulated by an increase in the blood concentration of sodium caused by dehydration. ADH increases water reabsorption in the kidney, thus eliminating the stimulus for ADH secretion.

The thyroid may malfunction, producing too little T_3 and T_4. This condition, known as **hypothyroidism**, results in a decrease in the metabolic rate. People suffering from it feel cold much of the time. They may also feel tired and worn out. Even simple mental tasks become difficult. Their heart rate may slow to 50 beats per minute. Hypothyroidism is treated by pills containing artificially produced thyroid hormone.

Excess thyroid activity, **hyperthyroidism**, in adults results in elevated metabolism. Patients suffer from excessive sweating (due to overheating). They may become thin, even if they eat a lot. The increase in thyroid hormone levels results in increased mental activity, resulting in nervousness and anxiety. People with hyperthyroidism often find it difficult to sleep. Their heart rate may accelerate, and they may lose their sensitivity to cold.

Patients may be given antithyroid medications, drugs that block the effects of thyroid hormones. Surgery may be required to remove part or all of the gland if it has become cancerous. The most common treatment for hyperthyroidism, however, is radioactive iodine. Iodine, a component of thyroxine and T_3, is concentrated in the thyroid gland where these hormones are made. Radioactive iodine therefore accumulates in the hormone-producing cells, damaging them and reducing their output of thyroid hormones.

Calcitonin. **Calcitonin** is produced by other cells of the thyroid gland. This hormone lowers the blood calcium level (Figure 13-10). It does so in part by stimulating the formation of new bone. It also inhibits the action of bone-destroying cells discussed in Module 11.

Calcitonin is involved in a simple negative feedback loop with calcium ions in the blood. When the calcium-ion concentration increases, calcitonin secretion increases. As calcium concentrations fall, calcitonin secretion falls.

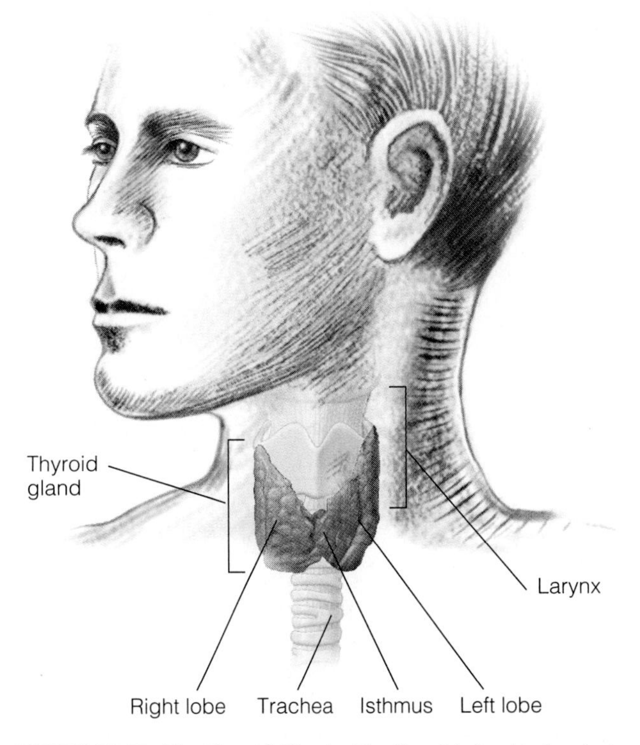

FIGURE 13-10 The Thyroid Gland The thyroid gland is located in the neck on either side of the larynx.

The Parathyroid Glands

The **parathyroid glands** are four small nodules of tissue embedded in the back side of the thyroid gland. These glands produce a hormone known as **parathyroid hormone** or **PTH**.

Parathyroid hormone has the opposite effect of calcitonin. That is, it increases blood calcium levels. Its secretion is stimulated when calcium levels in the blood drop. PTH restores blood calcium levels in one of several ways. For example, it can increase the amount of calcium absorbed in the small intestine. It can also stimulate bone destruction.

Calcium levels are also influenced by vitamin D, available in some foods and beverages such as milk. Vitamin D is also produced in the skin when it is exposed to sunlight. Vitamin D increases calcium absorption in the intestine. In addition, it increases the responsiveness of bone to parathyroid hormone.

As in other glands, the parathyroids may malfunction. Excess secretion of parathyroid hormone, the most common condition, may result from a tumor in the parathyroid gland that causes the secretion of excess PTH. Excess PTH, in turn, results in a loss of calcium from the bones and teeth. Because bones contain enormous amounts of calcium, most symptoms do not appear until 2-3 years after the onset of the disease. Therefore, by the time the disease is discovered, kidney stones may already have formed from calcium, cholesterol, and other substances. Bones may have become more fragile and susceptible to breakage. To prevent further complications, parathyroid tumors must be removed.

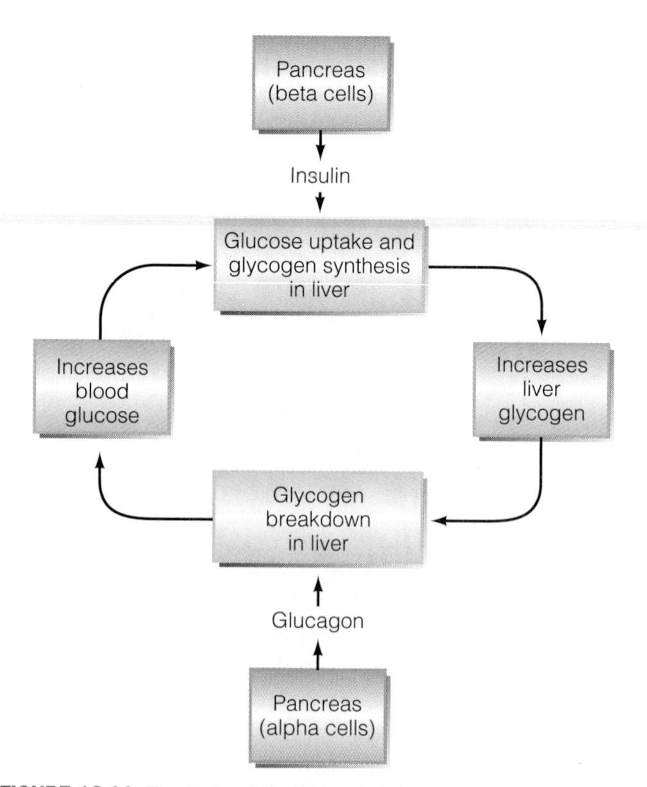

FIGURE 13-11 The Role of the Liver and Pancreas in Controlling Blood Glucose Levels Glucagon and insulin are antagonistic hormones that regulate blood glucose levels through different mechanisms.

The Pancreas

Insulin is a hormone produced by the pancreas, an organ that also produces digestive enzymes (Module 3). Insulin has many functions but one of the most important is that it stimulates the uptake of glucose by body cells. It also stimulates the synthesis of glycogen in liver and muscle cells. As noted in Module 3, glycogen is a molecule made up of many glucose molecules. It's the way the body stores glucose for use between meals.

In healthy individuals, the amount of glycogen in the liver increases immediately after meals. It then declines in the period between meals as the body uses glucose. Muscle glycogen also increases after eating, then declines as the muscles break it down to produce glucose needed to provide energy.

Insulin also affects fat storage. When present, insulin stimulates lipid synthesis, creating food stores for times of need.

The pancreas also produces a hormone known as *glucagon*. **Glucagon** has the opposite effect of insulin. That is, it increases blood levels of glucose by stimulating the breakdown of glycogen in liver cells (**FIGURE 13-11**). This process helps to maintain proper glucose levels in the blood between meals.

Glucagon secretion, like that of insulin, is regulated by a negative feedback mechanism controlled by glucose concentrations in the blood. When glucose levels fall, the pancreas releases glucagon into the bloodstream. When glucose levels rise, glucagon secretion declines.

Diabetes Mellitus. Most readers have heard about a disease called *diabetes* or *diabetes mellitus*. Physicians recognize two types of diabetes. **Type I diabetes** begins early in life and is also called **early-onset diabetes.** This disease is believed to be caused by damage to the insulin-producing cells of the pancreas. In some individuals diabetes is the result of an autoimmune reaction. Others may develop it as a result of a viral infection or an environmental pollutant that damages the cells that produce insulin. Insulin production varies in patients with Type I diabetes. In some, it is only slightly reduced; in others, it is completely suppressed.

Type II diabetes usually occurs in people over 40 and is also called **late-onset diabetes**. In this disease, the cells of the pancreas produce normal or above-normal levels of insulin. However, the target cells of the body are unresponsive to the hormone.

Type II diabetes is commonly associated with obesity, a problem growing rapidly in the United States where people consume high-fat, high-calorie diets. Because obesity in adults and children is on the rise, Type II diabetes is also increasing very rapidly.

Although Type I and Type II diabetes have different causes, both forms of this disease exhibit similar symptoms. Excess urination and thirst are generally the first signs of trouble. Patients often feel tired, weak, and apathetic. Weight loss and blurred vision are also common. Excess glucose in the urine may result in frequent bacterial infections in the bladder.

Early-onset diabetes (Type I) is treated with insulin injections. Patients give themselves regular injections of insulin—usually two to three times per day. Patients are also required to eat meals and snacks at regular intervals to maintain constant glucose levels in the blood and to ensure that regular insulin injections always act on approximately the same amount of blood glucose.

To mimic the body's natural release, medical researchers have developed a device called an *insulin pump*. This device delivers predetermined amounts of insulin. Researchers are also experimenting with ways to transplant healthy insulin-producing cells in the pancreases of diabetics.

Although insulin injections help in treating Type I diabetes, they're utterly useless in treating Type II, or late-onset diabetes. The reason for this is that this disease very likely stems from a lack of insulin receptors in target cells.

In many patients, Type II diabetes can be eliminated by weight loss. In others, the disease can be controlled by diet and exercise. For example, physicians restrict carbohydrate intake of their patients and instruct them to eat small meals at regular intervals during the day. Candy, sugar, cakes, and pies are strictly off-limits.

Although treatments for both kinds of diabetes have improved people's lives, risks are still present. Type I diabetics, for example, may suffer from diabetic comas, or unconsciousness. This occurs when they receive insufficient amounts of insulin or if they skip an insulin injection or two. Without insulin, the body cells become starved for glucose (even though blood levels are high) and begin breaking down fat. Excessive fat catabolism releases toxic chemicals (called ketones) that cause the patient to lose consciousness.

Diabetics may also suffer from **insulin shock** caused by an overdose of insulin. This reduces blood glucose levels, creating hypoglycemia. In mild cases, symptoms include tremor, fatigue, sleepiness, and the inability to concentrate. These symptoms result from a lack of glucose in the brain. In severe cases, unconsciousness and death may occur.

Over the long term, patients with both forms of diabetes may experience more damaging effects. For example, some individuals suffer from loss of vision, nerve damage, and kidney failure. These symptoms appear 20-30 years after the onset of the disease, even if they are being treated. Damage to the circulatory system can cause gangrene (GANG-green), requiring amputation of limbs, especially the lower extremities. These serious complications result from the inevitable elevations in blood glucose levels that occur periodically over the years. Elevated levels of glucose can damage nerve cells and blood vessels.

The Adrenal Glands

Atop the kidneys are two endocrine organs, the adrenal glands. The **adrenal glands** (ah-DREE-nal), shown in **FIGURE 13-12**, consists of two zones. The central region, or **adrenal medulla**, produces the hormones that increase the heart rate and accelerate breathing when a person is excited or frightened. The outer zone, the **adrenal cortex**, produces a number of steroid hormones, discussed shortly.

Hormones of the Adrenal Medulla. The adrenal medulla produces two hormones: **adrenalin** (epinephrine) and **noradrenalin** (norepinephrine). In humans, about 80% of the adrenal medulla's

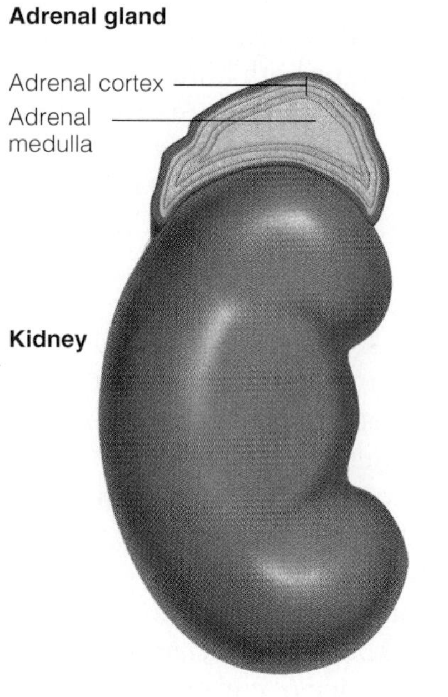

FIGURE 13-12
Adrenal Gland The adrenal glands sit atop the kidney and consist of an outer zone of cells, the adrenal cortex, which produces a variety of steroid hormones, and an inner zone, the adrenal medulla. The adrenal medulla produces adrenalin and noradrenalin.

output is adrenalin. Helping us meet the stresses of life, adrenalin and noradrenalin are instrumental in the **fight-or-flight response**—the physiological reactions that take place when an animal is threatened. These hormones enhance our ability to either fight or flee.

Adrenalin and noradrenalin are secreted under stress—for example, when a careless driver cuts in front of you in traffic or as you wait outside a lecture hall to take an exam. Nerve impulses traveling from the brain to the adrenal medulla trigger the release of adrenalin and noradrenalin.

The changes these hormones cause are many. For example, they elevate blood glucose levels, making more energy available to cells, particularly skeletal muscle cells. They also increase breathing rate and heart rate. In addition, these hormones cause tiny air-carrying tubules in the lungs to dilate. This permits greater movement of air in and out of them. Furthermore, these hormones cause blood vessels in the intestinal tract to constrict, putting digestion on temporary hold. This diverts blood to the dilated blood vessels in skeletal muscles increasing flow through them. Mental alertness increases as a result of increased blood flow and hormonal stimulation. You are ready to fight or flee.

Hormones of the Adrenal Cortex. The adrenal cortex produces two main types of hormones, each of which has a different function. The first group, the glucocorticoids, helps to maintain blood glucose levels. Several chemically distinct glucocorticoids are secreted, the most important being cortisol.

The second group, the **mineralocorticoids** (MIN-er-al-oh-CORE-teh-KOIDS), regulate the concentration of ions in the blood and tissue fluids. The mineralocorticoids are involved in maintaining the proper concentration of certain ions, notably potassium and sodium. The most important mineralocorticoid is **aldosterone**. It acts on the kidney as explained in Module 8.

Diseases of the Adrenal Glands. Like other endocrine glands, the adrenal can run amok. One of the most common disorders is **Addison's disease**. Most cases of Addison's disease are thought to be autoimmune reactions in which cells of the adrenal cortex are destroyed by the immune system.

Addison's disease results in a decrease in the production and release of hormones from the adrenal cortex. The absence of cortisol upsets the body's homeostatic mechanism for controlling glucose. The lack of aldosterone results in low levels of sodium and low blood pressure. Symptoms include loss of appetite, weight loss, fatigue, and weakness.

Addison's disease can be treated with steroid tablets that replace the missing hormones. Treatment allows patients to lead fairly normal, healthy lives.

As you can see, the endocrine system plays a very important role in maintaining the many functions of the body. Working in concert with the nervous system, it helps ensure our day-to-day functions and our long-term survival.

The Male Reproductive System

Most body systems you've studied so far function to keep us alive and well. The reproductive system is not necessary for individual health and survival. Rather, it functions to keep our species alive. In this module, we'll begin our exploration of human reproduction, starting with the male reproductive system.

Overview of the Male Reproductive System

The male reproductive system consists of a number of organs. The **penis** is the organ of copulation. Below it, suspended in a sac of skin is the **scrotum**. It contains the **testes** (TESS-teas) or testicles. They produce sperm and male hormones. As you will soon see, the male reproductive system also consists of several organs that produce part of the semen, a place to store sperm, and some tubules that transport the sperm to and through the penis.

The Testes

Although the testes reside in the scrotum, they are actually formed inside the body cavity during fetal development. After formation, the testes descend into the scrotum. In a very small percentage of boys, however, the testes fail to descend. Drugs may be given to stimulate the testes to descend. If drug treatment is unsuccessful, surgery is required. If the condition is not corrected, sterility may result because the temperature inside the body cavity is too high for sperm development. The scrotum provides just the right temperature to permit normal sperm production.

As **FIGURE 14-1** shows, each testis is surrounded by a dense layer of connective tissue. This layer contains numerous pain fibers, a fact to which most men will attest. Each testis contains numerous tubules, in

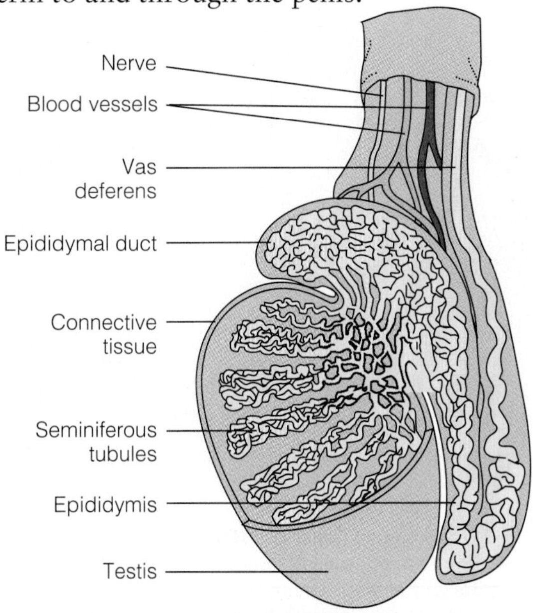

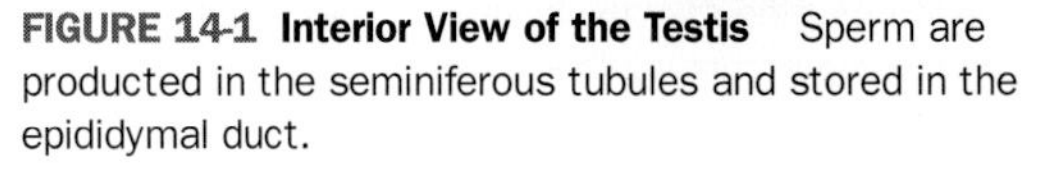

FIGURE 14-1 Interior View of the Testis Sperm are producted in the seminiferous tubules and stored in the epididymal duct.

which sperm are formed. They are called the **seminiferous tubules** (SEM-in-IF-er-uss).

Sperm produced in these tubules empty into a network of connecting tubules in the "back" of the testes. These tubules, in turn, empty in a duct where they are stored until released during **ejaculation**.

During ejaculation, the sperm from each testes enter a much larger, muscular duct known as the **vas deferens** (DEAF-er-en-she-ah). These ducts pass from the scrotum into the body cavity through two canals in the body wall, known as the **inguinal canals** (FIGURE 14-2A). As shown in FIGURE 14-3, each vas deferens empties into the urethra. It carries the sperm through the penis to the outside.

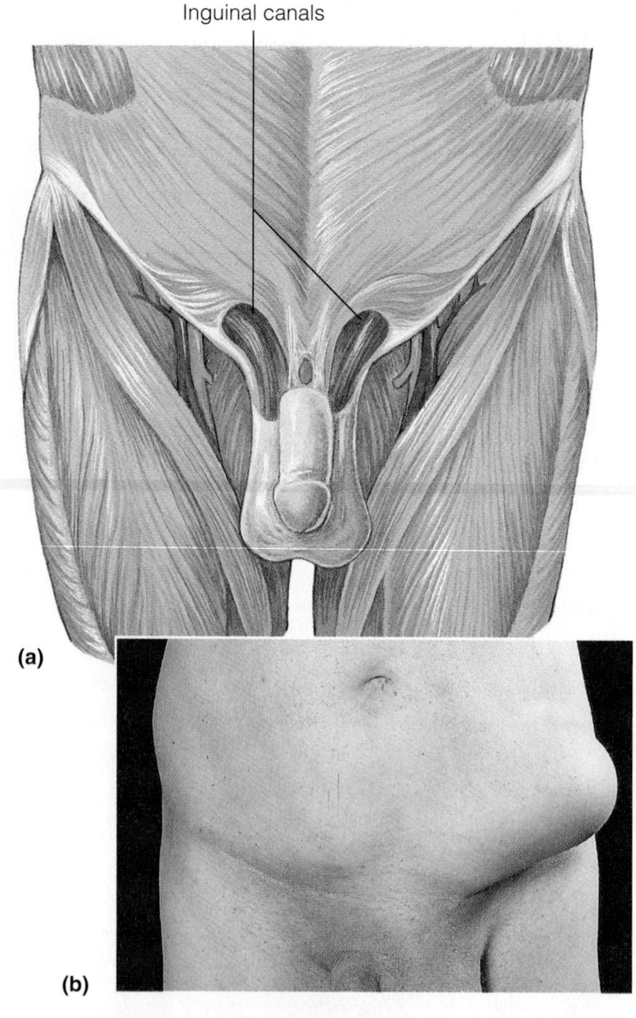

Inguinal canals

(a)

(b)

FIGURE 14-2 The Inguinal Canal and Hernia *(a)* During development, the testis descends through the inguinal canal, an opening through the musculature in the lower abdominal wall. *(b)* Loops of intestine may push through the inguinal canal if the muscles surrounding it are weak.

The inguinal canals are potential weak spots in the lower abdominal wall. In some men, parts of the small intestine may enter the weakened canals (FIGURE 14-2B). This condition is known as an **inguinal hernia** and can be corrected surgically. Weaknesses in the canal can be detected by a doctor who places a finger over the inguinal canal, then asks a man to cough. If the canal is weak, coughing will push the intestines into the canal.

During ejaculation, sperm are joined by fluids produced by additional glands, known as the *sex accessory glands*. They are located near the neck of the urinary bladder, (Figure 14-3). The sex accessory glands produce fluid that makes up 99% of the volume of the **ejaculate**, or **semen** (SEA-men). The remaining 1% consists of sperm produced in the testes.

One of the most prominent sex accessory glands is the **prostate gland** (PROS-tate). The prostate surrounds the neck of the bladder and empties its contents directly into the urethra. Routine medical examinations of men over the age of 45 show that nearly all of them have enlarged prostates. This condition results from the formation of small nodules inside the gland. Although they usually cause no trouble, in some cases the nodules grow quite large and can block the flow of urine, making urination painful.

The prostate is also a common site for cancer in men over 40 and should be checked every year by a physician once men reach that age. Prostate cancer typically remains undetected until the tumor begins to press on the urethra, making urination difficult and painful. Prostate cancers are detected by digital rectal exams that allow doctors to feel the prostate. A blood test is also used to determine cancer.

If cancer is confined to the gland, it can be removed surgically. Radiation can also be used to treat prostate cancer. If the cancer has spread, however, it may be too late. Drugs that inhibit testosterone's secretion or its actions can be given to slow the growth of the tumors. The testes may also have to be removed to reduce testosterone secretion.

On average, men produce 200-300 million sperm every day. The average ejaculate contains 240 million or more. In humans, each sperm formed during meiosis contains 23 single-stranded chromosomes—half the number in a normal body cell. Thus, when the sperm unites with an egg (also

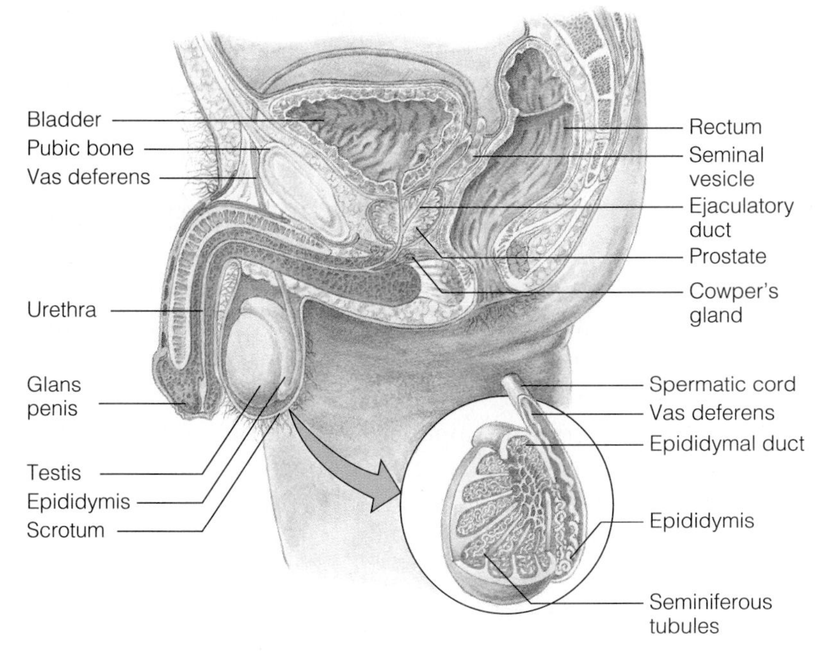

FIGURE 14-3 Anatomy of the Male Reproductive System

containing 23 single-stranded chromosomes), they produce a cell containing 46 single-stranded chromosomes. One-half of its chromosomes come from each parent.

The testes also produce male sex hormones. These hormones are produced by cells found in between the seminiferous tubules. These cells produce a group of sex steroid hormones known as **androgens**, so named because they have a masculinizing effect. The most important androgen is **testosterone**.

Testosterone has many functions. It stimulates the formation of sperm. In the absence of testosterone, sperm cell production declines, then stops, and the walls of the seminiferous tubules shrink.

Testosterone stimulates growth in bone and muscle and explains in part why men are generally taller and more massive than women. In addition, testosterone promotes facial hair growth and thickening of the vocal cords, giving men deeper voices than women.

Testosterone affects the hair follicles on the heads of many men, causing baldness. It is not the absence of testosterone, as some believe, but the presence of testosterone and certain genes that lead to this condition.

Testosterone also stimulates the oil glands of the skin in both sexes (**FIGURE 14-4A**). During puberty (sexual maturation) in boys, testosterone levels rise dramatically. This causes a sharp increase in oil gland activity. Dead skin cells may block the pores that open onto the skin's surface (**FIGURE 14-4B**). As a result, oil collects inside the glands. Bacteria on the skin often invade and proliferate in the small pools of oil. This causes inflammation, pus formation, and swelling. The result is an **acne pimple**.

Mild acne can be treated by washing the skin twice a day with warm water and a mild, unscented soap. Be sure to wash gently. You should also apply a cream containing benzoyl peroxide daily to affected areas. Women should avoid makeup that has an oil base or use a nonoily type of foundation and wash their faces thoroughly each night. Sunlight also helps clear up acne, because it dries the oil on the skin and kills skin bacteria. Doctors may prescribe antibiotic creams. Pimples should not be picked at either, as this may worsen acne and cause scarring.

Moderate acne can be successfully treated with antibiotics. Severe acne can be treated by special ointments and antibiotics given by a skin doctor. One relatively new and fairly successful drug is Retin-A, a derivative of vitamin A.

The Penis

Sperm are deposited in the female reproductive tract with the aid of the penis. The penis consists of a shaft of varying length and an enlarged tip (**FIGURE 14-5**). The tip is covered by a sheath of skin at birth, the foreskin. The foreskin gradually becomes separated from the glands in the first two years of life. At puberty, the inner lining of the foreskin begins to produce an oily secretion. Bacteria can grow in the protected, nutrient-rich environment created by the foreskin, so special precautions must be taken to keep the area clean.

Because of potential health problems or religious reasons, many parents opt to have the foreskin removed in the first few days of their son's life. The operation, called **circumcision** (sir-come-SIZH-un; literally, "to cut around"), may help reduce penile cancer in men and may also reduce cervical cancer in the wives or sexual partners of circumcised men.

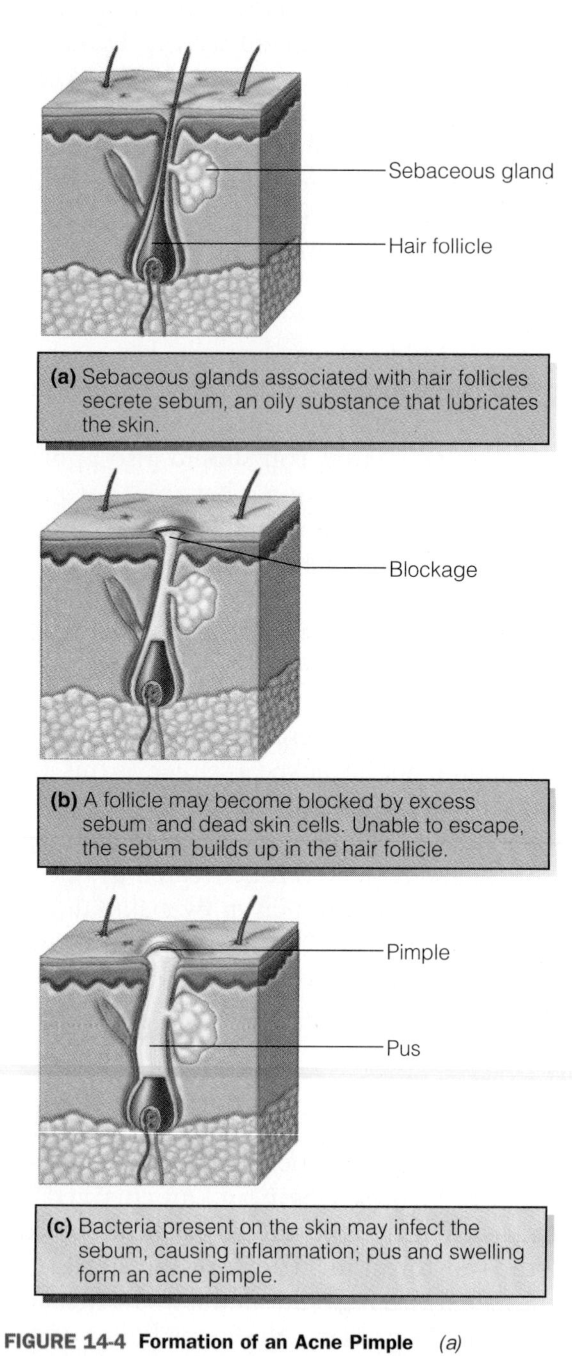

FIGURE 14-4 Formation of an Acne Pimple *(a)* Testosterone stimulates oil production in the sebaceous (oil) glands. *(b)* If the outlet is blocked, sebum builds up in the gland and *(c)* the gland may become infected.

For successful copulation, the penis must become rigid, or erect. During sexual arousal, nerve impulses cause arterioles in the penis to dilate. Blood flows into a spongy **erectile tissue** (eh-REK-tile) in the shaft of the penis, making it harden. Swelling compresses a large vein on the dorsal surface of the penis, blocking the outflow of blood and further stiffening the organ.

In the center of the penis is the urethra, a duct that carries urine from the bladder to the outside of the body during urination. The urethra also transports semen—sperm and secretions of the sex accessory glands—during ejaculation.

Some men lose their ability to achieve or sustain an erection. This condition is known as **impotence** (IM-poe-tense). Most men experience impotence at some time in their life, but it is usually temporary. Persistent impotence, however, is a more serious condition, and is more common in middle-age and elderly men.

Persistent impotence may be caused by a number of factors. Marital conflict, stress, fatigue, and anxiety, for example, all contribute to impotence. Nerve damage may also cause impotence. Alcohol ingestion and some medications may contribute to the condition as well.

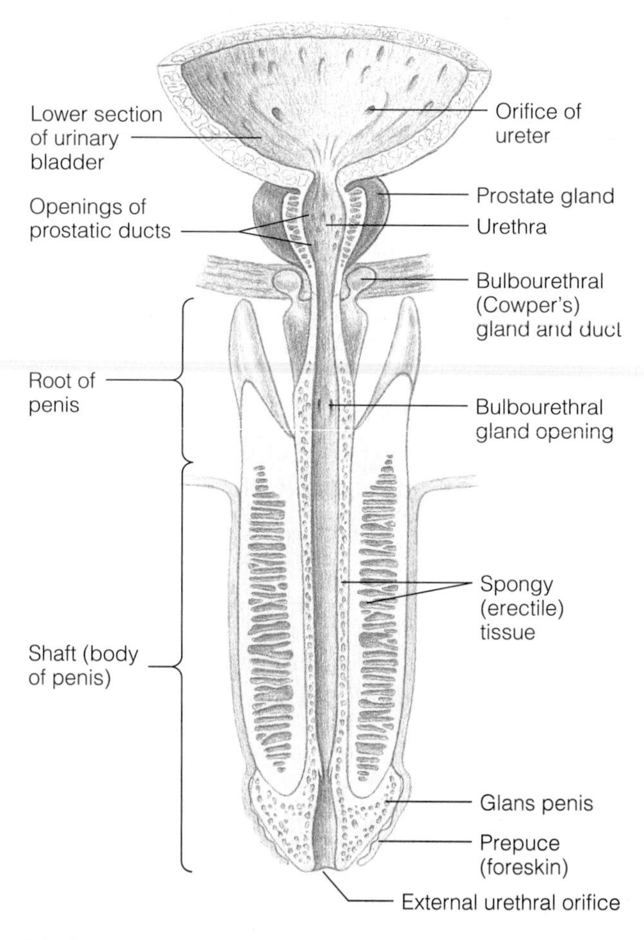

FIGURE 14-5 Anatomy of the Penis The penis consists principally of spongy tissue that fills with blood during sexual arousal. The urethra passes through the penis, carrying urine or semen.

Smoking is also a major contributor to impotence. In fact, smokers are twice as likely to be impotent as nonsmokers. That's because smoking causes arterioles in the penis to constrict, reducing blood flow. Smoking also causes the buildup of plaque, which permanently blocks blood flow to this organ.

If the problem is psychological, therapy is often advised. Rest and relaxation help. If the problem is a result of high cholesterol that causes plaque build up in the arteries supplying the penis, surgery can be helpful. If the problem is a medication one is taking, a change of prescription can help. Patients with nerve damage, however, are likely to suffer permanent impotence.

Nerve damage may result from diabetes or from an accident. For patients with irreversible impotence, urologists can surgically insert an inflatable plastic implant in the penis. The implant is attached to a small, fluid-filled reservoir in the scrotum. The fluid is manually pumped into the implant, making the penis erect upon demand, thus permitting sexual intercourse. Other types of implants are also available. Vacuum pumps can be used, too. These devices use a plastic cylinder that is fitted over the penis. A hand-held pump is used to create a vacuum. Blood rushes into the penis, causing an erection. After the vacuum aspirator is removed, special rubber bands are placed around the base of the penis to maintain an erection.

A number of drugs are also available. Viagra, for example, promotes erection when men are sexually aroused. This drug works by causing the muscle in the walls of the arteries supplying the penis to relax. This permits blood to flow into the organ more readily when stimulated.

Ejaculation

Ejaculation is a reflex involving the nervous system. Sexual stimulation results in nerve impulses in sensory nerves to the spinal cord. When stimulation becomes intense, these impulses activate neurons in the cord. These neurons, in turn, send impulses to the smooth muscle in the walls of the tubes that store sperm in each testes and to the sex accessory glands and the vas deferens. Sperm and secretions of the sex accessory glands are released and propelled to the urethra. Semen is then propelled onward by smooth muscle contractions in the walls of the urethra and is released in spurts.

FIGURE 14-6 Hormonal Control of Testicular Function
Testosterone, FSH, and ICSH participate in a negative feedback loop. The testes also produce a substance called *inhibin*, which controls GnRH secretion.

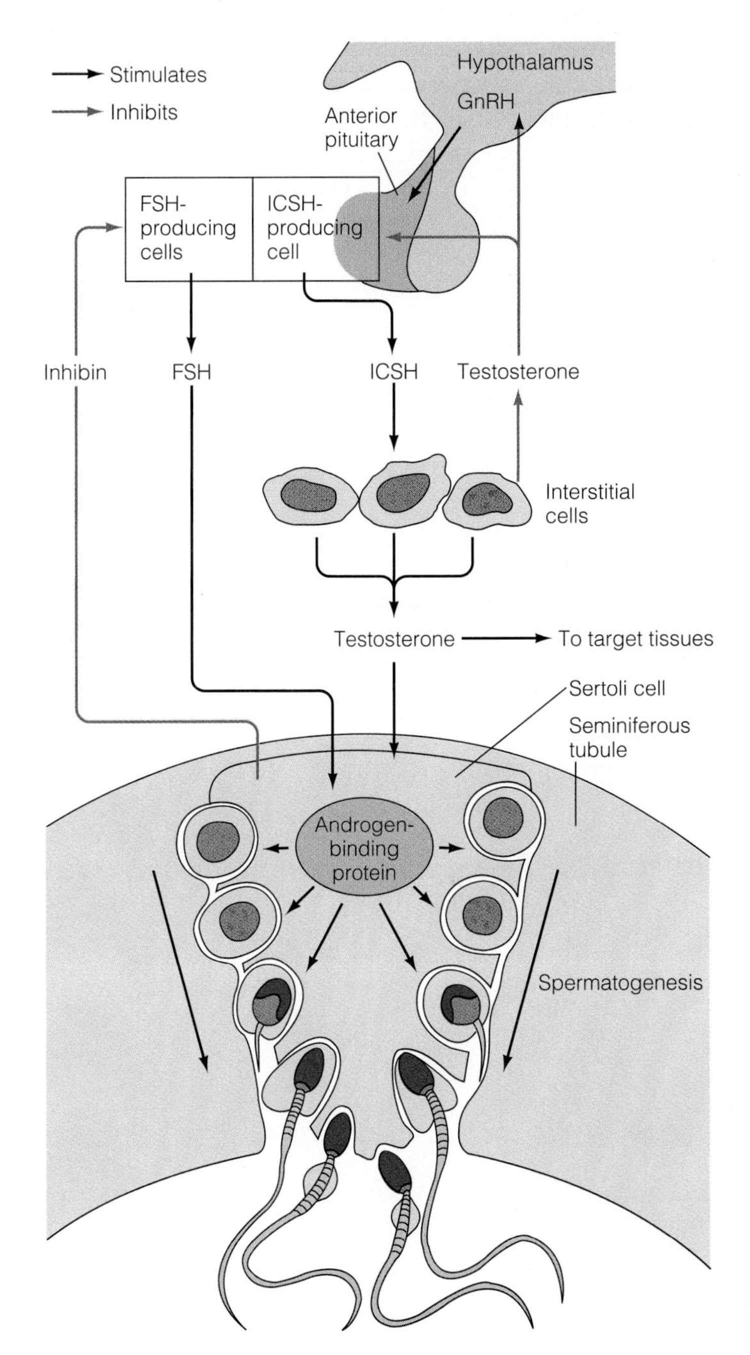

Control of the Male Reproductive System

As noted earlier, the testes produce sex steroid hormones—mainly, testosterone. Testosterone production and release, however, are controlled by a hormone from the pituitary gland. This hormone is known as **luteinizing hormone** (**LH**). In males, LH is also known as **interstitial cell**

stimulating hormone (**ICSH**). It gets this name because it stimulates the testosterone-producing cells (interstitial cells).

ICSH secretion is controlled by a releasing hormone produced by the hypothalamus. That hormone is known as **gonadotropin releasing hormone** (**GnRH**). As **FIGURE 14-6** shows, the secretion of GnRH is controlled by testosterone levels in the blood in a classic negative feedback loop. When testosterone levels in the blood decline, receptors in the hypothalamus detect the change and signal an increase in GnRH secretion.

The pituitary also produces the gonadotropin **follicle-stimulating hormone** or **FSH**. Like testosterone, FSH stimulates sperm formation (Figure 14-6).

Human reproduction is an intricate function. Although it is not essential to our own lives, it does provide us with the means to continue the life of our species.

The Female Reproductive System

The female reproductive system like the male reproductive system plays a key role in maintaining our species. As you will soon see, the female reproductive system consists of the external genitalia and several internal organs. We'll start with the internal structures: (1) the ovaries, (2) the uterine tubes, (3) the uterus, and (4) the vagina.

The Uterus and Uterine Tubes

The **uterus** or **womb** is a pear-shaped organ about 7 centimeters (3 inches) long and about 2 centimeters wide (less than 1 inch) at its broadest point in nonpregnant women. The wall of the uterus contains a thick layer of smooth muscle. The uterus houses and nourishes the developing fetus.

Attached to the uterus are two hollow, muscular tubes, known as the **uterine tubes** (YOU-ter-in). In humans, the uterine tubes are often referred to as the **Fallopian tubes** (fal-OH-PE-ee-an). As **FIGURE 15-1A** shows, the ends of the uterine tubes are widened like a catcher's mitt and fit loosely over the ovaries. The **ovaries** (OH-var-ees) are paired, almond-shaped organs.

Fertilization—the union of the sperm and egg—occurs in the upper third of the uterine tubes. The fertilized egg is then transported down the uterine tubes to the uterus. Inside the uterus, it attaches to the lining, known as the *endometrium* (EN-doh-MEE-tree-um). It embeds itself here, remaining for the duration of pregnancy.

At birth, the baby is expelled from the uterus through the **cervix** (SIR-vix), the lowermost portion of the uterus. As **FIGURE 15-1** shows, the cervix protrudes into the vagina (vah-GINE-ah). Although the canal running through the cervix is quite narrow, the cervix stretches considerably at birth.

The **vagina** or **birth canal** is a 3-inch, tubular organ that leads to the outside of the body. The vagina also serves as the receptacle for sperm during sexual intercourse. To reach the ovum, sperm must travel through a tiny opening and narrow canal of the cervix that leads into the uterus. From here, sperm move up both uterine tubes.

The **external genitalia** are the externally visible parts of the female reproductive system. They consist of two flaps of skin on either side of the vaginal opening (Figure 15-1). The

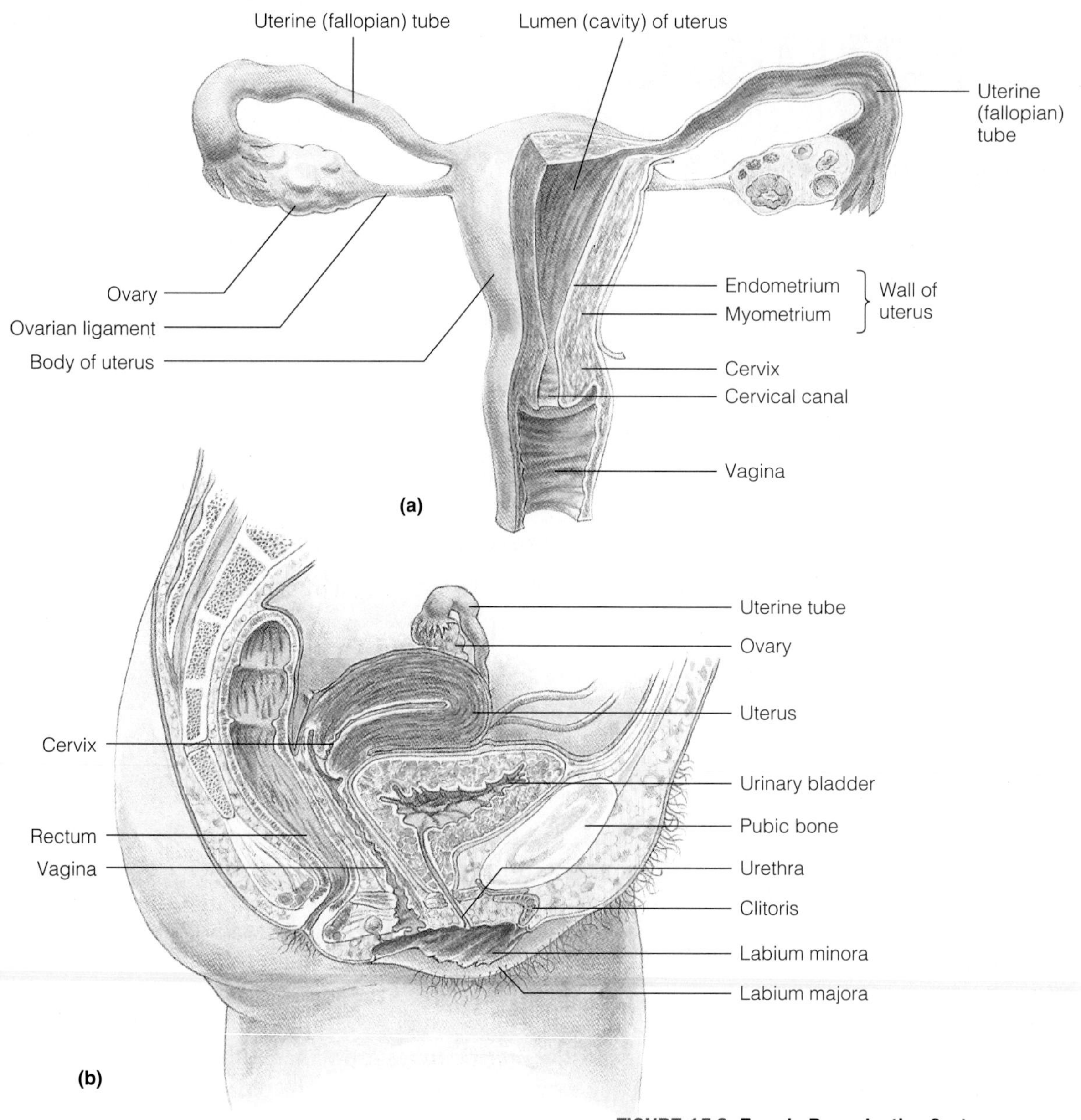

FIGURE 15-1 Female Reproductive System
(a) Frontal view. *(b)* Midsagittal view.

outer folds are the **labia majora** (LAY-bee-ah ma-JOR-ah). These large folds of skin are covered with hair. The inner flaps are the **labia minora** (meh-NOR-ah). They meet to form a hood over a small knot of tissue called the clitoris (CLIT-er-iss). The **clitoris** is a highly sensitive organ involved in female sexual arousal. It is formed from the same embryonic tissue as the penis. The various components of the female reproductive system, and their functions, are listed in **FIGURE 15-2**.

FIGURE 15-2 Female Reproductive System

The Female Reproductive System

Component	Function
Ovaries	Produce ova and female sex steroids
Uterine tubes	Transport sperm to ova; transport fertilized ova to uterus
Uterus	Nourishes and protects embryo and fetus
Vagina	Site of sperm deposition, birth canal

The Ovaries

During each menstrual cycle, one ovary releases an egg or **ovum**, the female gamete. This process is called *ovulation* (OV-you-LAY-shun). The ovum or egg is drawn into the uterine tube.

The release of an egg occurs approximately once a month in women during their reproductive years—from puberty (age 11-15) to menopause (age 45-55). Ovulation is temporarily halted when a woman is pregnant.

The structure of an ovary is shown in **FIGURE 15-3**. Eggs are surrounded by one or more layers of cells derived from the loose connective tissue of the ovary. Together, the eggs and surrounding cells form follicles.

During each cycle about a dozen follicles begin to enlarge, as shown in Figure 15-3. As they get bigger, a clear liquid begins to accumulate between the follicle cells. The fluid creates small spaces among the follicle cells, which enlarge as additional fluid is generated. Eventually, the cavities join, forming one large cavity.

Although a dozen or so follicles begin developing during each cycle, as a rule only one makes it to ovulation. The rest stop growing and degenerate. The follicle that survives continues to enlarge by accumulating more fluid. As the fluid builds up, the follicle begins to bulge from the surface of the ovary, not unlike a pimple. This weakens the wall. Eventually, the wall of the follicle breaks down, and the egg is released. The follicle collapses.

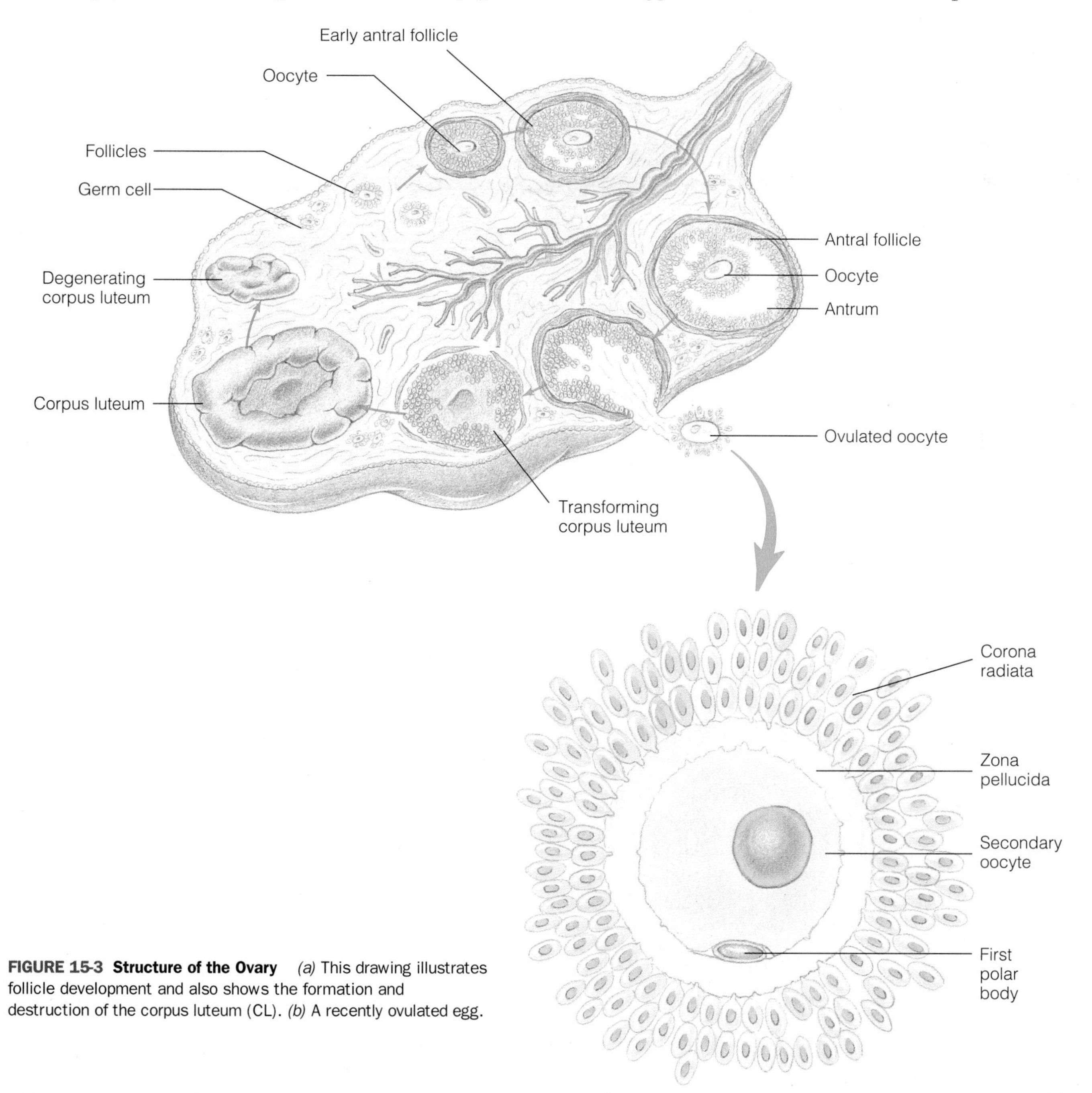

FIGURE 15-3 Structure of the Ovary *(a)* This drawing illustrates follicle development and also shows the formation and destruction of the corpus luteum (CL). *(b)* A recently ovulated egg.

The collapsed follicle does not degenerate. Rather, it forms a structure known as the **corpus luteum** (CORE-puss LEU-tee-um; "yellow body") or **CL** for short—so named because of the yellow pigment it contains in cows and pigs (Figure 15-3).

The CL is a temporary endocrine gland. It produces two sex hormones, estrogen and progesterone. If the egg is fertilized, the CL remains active for several months, producing hormones needed for a successful pregnancy. If fertilization does not occur, the CL soon disappears.

The Menstrual Cycle

Women of reproductive age undergo a series of changes called the **menstrual cycle** (MEN-strell). This cycle lasts from 25 days to 35 days, although the length of the cycle may also vary from month to month in the same woman. On average, however, the cycle repeats itself every 28 days. Ovulation occurs approximately at the midpoint of the cycle.

The menstrual cycle involves changes in the

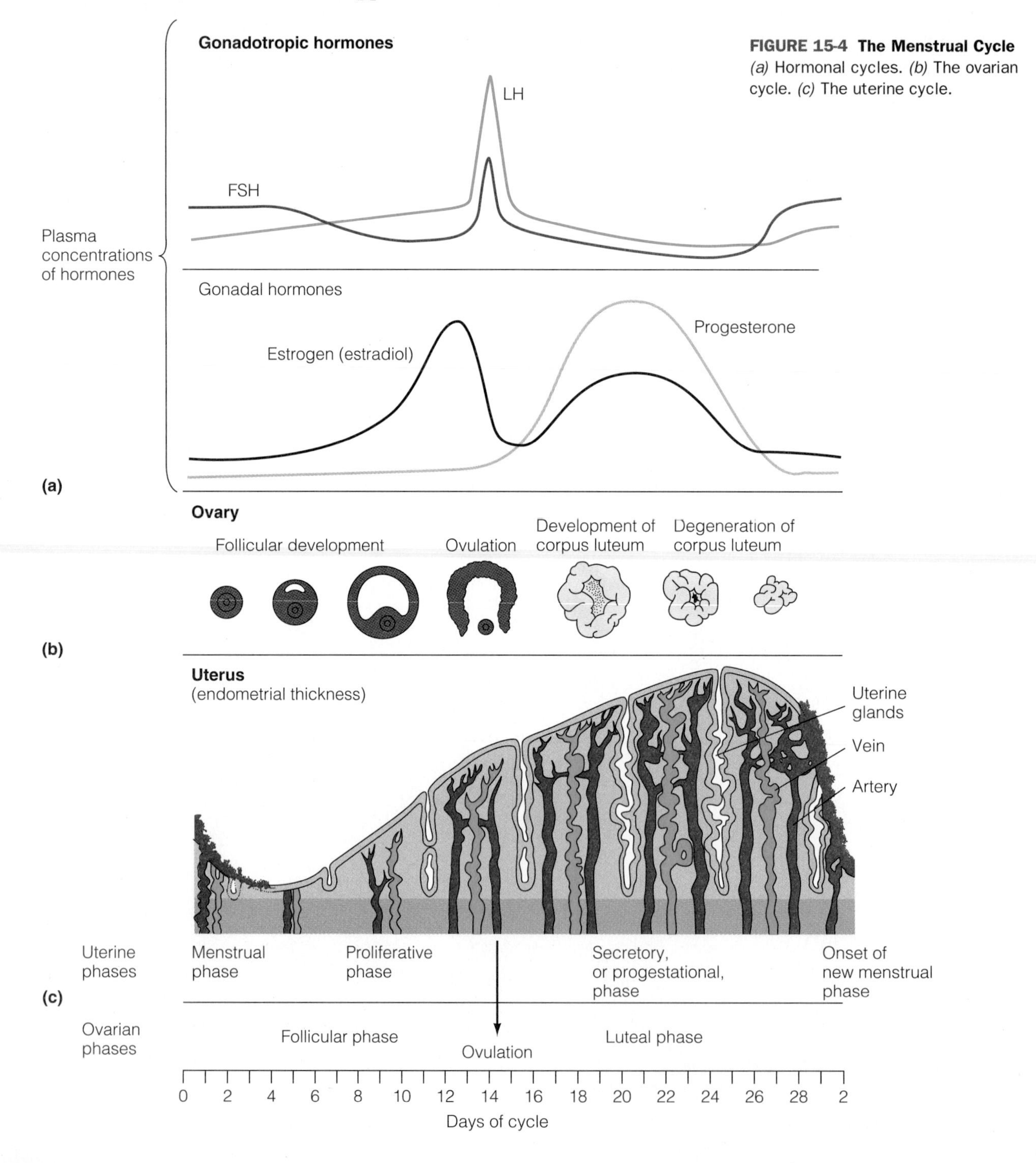

FIGURE 15-4 The Menstrual Cycle *(a)* Hormonal cycles. *(b)* The ovarian cycle. *(c)* The uterine cycle.

ovary and the uterus. These are brought about by changes in hormones. As shown in **FIGURE 15-4**, estrogen concentrations in the blood increase during the first half of the menstrual cycle. Estrogen production is stimulated by LH, a hormone from the pituitary.

Estrogen causes the lining of the uterus to increase in size, in preparation for a possible pregnancy. It also stimulates follicles to grow. Follicle growth is also stimulated by a pituitary hormone, **FSH**, short for **Follicle Stimulating Hormone**.

In the middle of each cycle, LH and FSH secretion peak. This surge in hormones causes ovulation.

During the second half of the menstrual cycle, the collapsed follicle forms the CL. It produces estrogen and progesterone. They cause the uterine lining to thicken even more in preparation for possible pregnancy. During this time, glands in the uterine lining swell with a glycogen-rich secretion.

If fertilization does not occur, the uterine lining starts to shrink approximately 4 days before the end of the cycle. It then begins to be shed, starting an event called **menstruation**. The shedding of the uterine lining (menstruation) is triggered by a decline in estrogen and progesterone concentrations in the blood.

When progesterone levels fall, the uterus begins to undergo contractions. These contractions propel the detached lining and blood from broken blood vessels out of the uterus into the vagina. These contractions cause the cramps that many women experience during menstruation.

If fertilization occurs, the newly formed embryo produces an LH-like hormone known as **HCG** (**human chorionic gonadotropin**). HCG stimulates the corpus luteum, maintaining its structure and function. When HCG is present, estrogen and progesterone continue to be secreted from the CL. The uterine lining remains intact. When the newly formed embryo arrives in the uterus, it attaches to the thickened lining. It then embeds in the lining from which it derives its nutrients. If the fertilized egg successfully embeds in the uterine lining, HCG will maintain the CL for approximately 6 months.

HCG shows up in detectable levels in a woman's blood and urine about 10 days after fertilization. Pregnancy tests available through a doctor's office or drugstore detect HCG in a woman's urine, and thus allow a women to determine if she is pregnant. The tests use a commercially prepared antibody to HCG, which binds to the hormone. The home pregnancy tests are relatively inexpensive, fairly reliable, and fast.

Estrogen and Progesterone

Like testosterone in boys, estrogen secretion in girls increases dramatically at puberty. As the levels of estrogen in the blood increase, the hormone begins to stimulate follicle development in the ovaries. Estrogen also stimulates the growth of the external genitalia and the breasts as well as internal structures, the uterus, uterine tubes, and vagina.

Estrogen's influence extends far beyond the reproductive system. For example, estrogen promotes rapid bone growth in the early teens. Because estrogen secretion in girls usually occurs earlier than testosterone secretion in boys, girls experience a growth spurt before similarly aged boys. However, estrogen also stimulates the closure of the growth zones of the bones. This puts an end to the female growth spurt fairly early. Thus, most girls reach their full adult height by the age of 15-17. Boys experience their most rapid growth later in adolescence and continue growing until the age of 19-21. Finally, estrogen stimulates the formation of fat in women's hips, buttocks, and breasts, giving the female body its characteristic shape.

Premenstrual Syndrome

Many women (4 out of 10) become irritable, depressed, and suffer from headaches and fatigue just before menstruation. Many also complain of bloating, tension, joint pain, and swelling and tenderness of the breasts. These complaints are symptomatic of a condition known as **PMS**, or **premenstrual syndrome**.

Physicians recommend that women suffering from PMS see their family doctor to be certain that the symptoms are not caused by some other medical problem. Doctors recommend relaxation and avoidance of stress to prevent or relieve PMS. Light exercise may help, as may warm baths. More frequent light meals with plenty of carbohydrates and fiber may help. Reductions in salt intake and avoiding excess chocolate consumption are also recommended. Cutting out caffeine drinks and taking vitamin B6 supplements are also generally advised.

Menopause

The menstrual cycle continues throughout the reproductive years. However, when a woman reaches 20, her ovaries gradually begin to become less responsive to FSH and LH. As a result, estrogen levels gradually decline. Ovulation and menstruation eventually stop. This end of these functions is called **menopause**—literally a cessation of the menses or menstruation.

Menopause generally occurs between the ages of 45 and 55, but can occur earlier. The decline in estrogen secretion causes the breasts and reproductive organs such as the uterus to begin to shrink. Vaginal secretions often decline, and, in some women, sexual intercourse becomes painful.

The decline in estrogen levels also results in behavioral changes. Many women, for instance, become more irritable and suffer periods of depression. Many women experience "hot flashes" and "night sweats" caused by massive dilation of vessels in the skin. Fortunately, these symptoms usually pass.

Declining estrogen levels also accelerate a softening of the bone due to a reduction in calcium. This is known as **osteoporosis**. To counter osteoporosis, physicians often prescribe small amounts of estrogen or, more commonly a combination of estrogen and progesterone. This treatment is called *hormone replacement therapy*. Doctors may also recommend a program of exercise and a diet rich in calcium and vitamin D as well. Most women are treated for 5 to 10 years after the onset of menopause, but not longer, for long-term treatment may increase the risk of breast cancer.

Birth Control

Birth control is any method or device that prevents births. Birth control measures fit into two broad categories: (1) **contraception**, ways of preventing pregnancy, and (2) **induced abortion**, the deliberate expulsion of a fetus.

FIGURE 15-5 summarizes the effectiveness of the most common means of contraception. A 95% effectiveness rating means that 95 women out of 100 using a certain method in a year will not become pregnant.

Not listed in the figure is a form of birth control known as **abstinence**, refraining from sexual intercourse. This method is appropriate for many people as a means of reducing unwanted pregnancy but has the added benefit of preventing the spread of AIDS and other sexually transmitted diseases (discussed later).

Sterilization is one of the most effective birth control measures. In women, sterilization is performed by cutting and tying off the uterine tubes, known as *tubal ligation* (TWO-bal lie-GAY-shun) (**FIGURE 15-6A**). This prevents sperm from reaching the eggs, and is very effective.

Male sterilization can be carried out in a physician's office under local anesthesia (**FIGURE 15-6B**). During this operation, called a *vasectomy*, the physician makes a small incision in the scrotum. Each vas deferens is then cut. The free ends are tied off or burned shut.

Vasectomies prevent the sperm from passing into the urethra during ejaculation. They do not decrease sex drive, and because they do not block fluids from the sex accessory glands, they have virtually no effect on ejaculation.

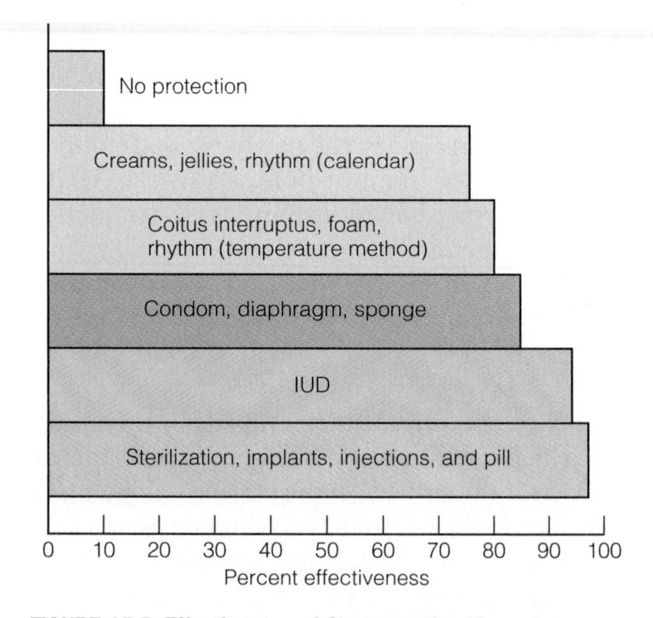

FIGURE 15-5 Effectiveness of Contraceptive Measures
Percent effectiveness is a measure of the number of women in a group of 100 who will not become pregnant in a year.

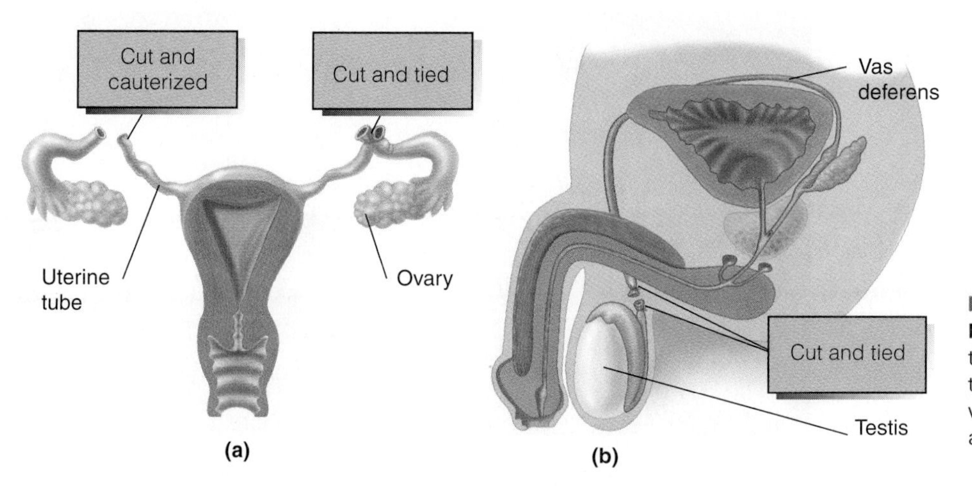

FIGURE 15-6 Sterilization Methods *(a)* In a tubal ligation, the uterine tubes are cut, then tied off or cauterized. *(b)* In a vasectomy, the vasa deferentia are cut, then tied off.

The **birth control pill** is also a highly effective means of birth control. The most commonly used birth control pill contains a mixture of estrogen and progesterone. These hormones inhibit the release of LH and FSH. This, in turn, inhibits follicle development and ovulation. Birth control pills must be taken throughout the menstrual cycle. Skipping a few days may result in ovulation and possible pregnancy.

Although effective, birth control pills have some adverse health effects. Although rare, some can be serious. One rare side effect is death. Death may result from heart attacks, strokes, or blood clots. The risk of a nonsmoker dying from taking birth control pills is 1 in 63,000 in any given year. The risk of dying in an auto accident is 1 in 6000.

Women who take birth control pills are more likely to develop cervical cancer than women who do not. Physicians therefore recommend annual Pap smears for women on the pill. During a Pap smear, the cervical lining is swabbed. The swab picks up cells that are examined under a microscope for signs of cancer.

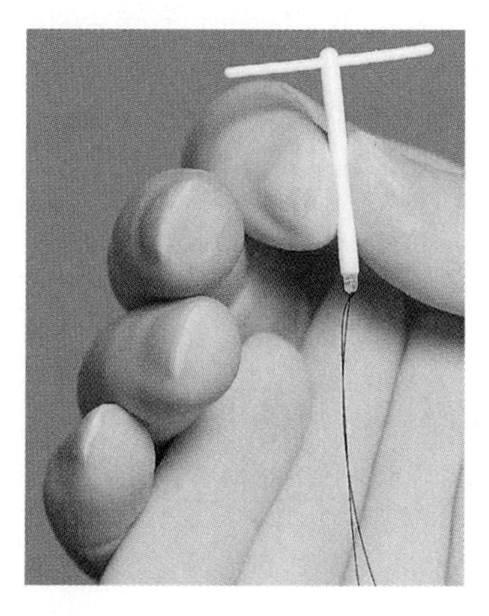

FIGURE 15-7 The IUD IUDs come in a variety of shapes and sizes and are inserted into the uterus, where they prevent implantation. Only one type is currently legal in the United States.

Smoking increases the likelihood of side effects from birth control pills. If a woman is a smoker and takes the pill, she is four times more likely to die from a heart attack or stroke than a nonsmoker. The risk of side effects also increases with age.

Birth control pills also have beneficial effects, not the least of which is that they prevent pregnancy. National statistics show that one out of every 10,000 women who becomes pregnant and delivers will die from complications, usually during delivery. Thus, even with the risks associated with the pill, using this mode of contraception is six times safer than pregnancy.

Birth control pills also reduce the incidence of a number of medical disorders including ovarian cysts, breast lumps, anemia, rheumatoid arthritis, osteoporosis, and pelvic infection. They may also protect a woman from cancer of the ovary and uterus.

The next most effective means of birth control is the **intrauterine device** (IUD) (**FIGURE 15-7**). The IUD is one of the least used birth control measures. It consists of a small plastic or metal object with a short string attached to it. IUDs are inserted into the uterus by a physician.

Like other forms of birth control, IUDs have adverse effects. In some cases, the uterus expels the device, leaving a woman unprotected. The IUD may also cause slight pain and increase menstrual bleeding. Although rare, IUDs can cause uterine infections and perforation (a penetration of the uterine wall by an IUD).

The next most effective means of birth control are the barrier methods—the diaphragm, condom, and vaginal sponge—all of which prevent the sperm from entering the uterus. The **diaphragm** (DIE-ah-

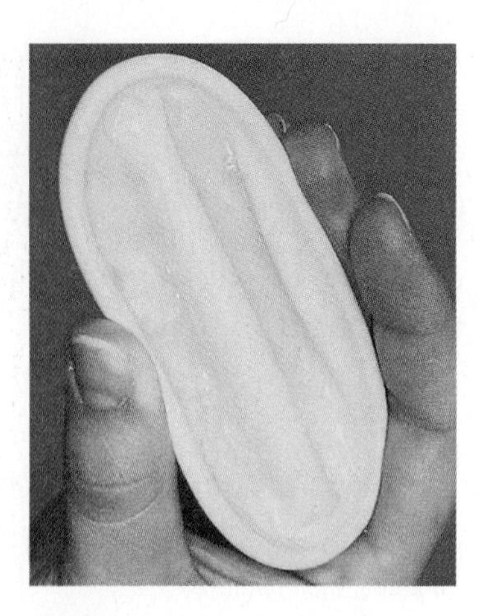

FIGURE 15-8 The Diaphragm Worn over the cervix, the diaphragm is coated with spermicidal jelly or cream and is an effective barrier to sperm.

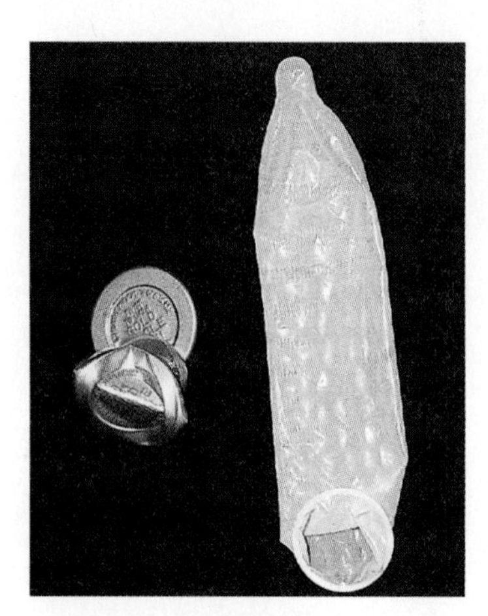

FIGURE 15-9 The Condom Worn over the penis during sexual intercourse, it prevents sperm from entering the vagina.

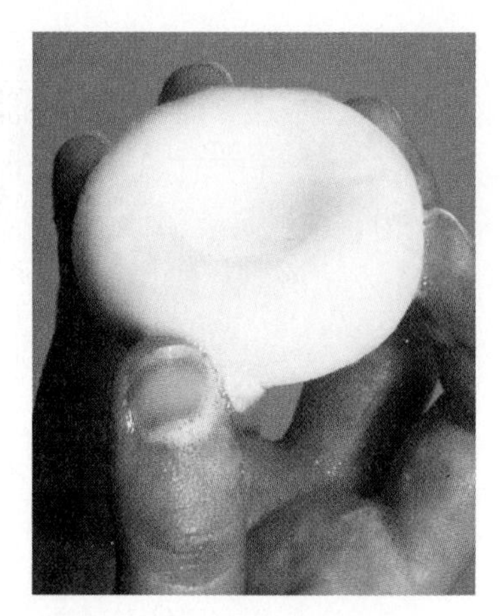

FIGURE 15-10 The Vaginal Sponge Impregnated with a spermicidal chemical, the vaginal sponge is inserted into the vagina and is effective for up to 24 hours.

FRAM) is a rubber cup that fits over the end of the cervix (FIGURE 15-8). To increase its effectiveness, a spermicidal (sperm-killing) jelly, foam, or cream is applied to the rim and inside surface of the cup.

Smaller versions of the diaphragm, called **cervical caps**, are also available. Fitting over the very end of the cervix, this device is held in place by suction. When used with spermicidal jelly or cream, the caps are as effective as full-sized diaphragms.

Condoms are thin, latex rubber sheaths that fit onto the erect penis (FIGURE 15-9). Sperm released during ejaculation are trapped inside. Besides preventing fertilization, condoms also protect against sexually transmitted diseases, a benefit not offered by any other birth control measure except abstinence.

Yet another barrier method is the **vaginal sponge** (FIGURE 15-10). This small absorbent piece of foam is impregnated with spermicidal jelly. Inserted into the vagina, the sponge is positioned over the end of the cervix. The sponge is effective immediately after placement and remains effective for 24 hours.

One of the oldest, but least successful means of birth control is **withdrawal**. The penis is removed just before ejaculation. This method requires tremendous willpower and frequently fails. Males may not pull out or may pull out too late. In addition, a few drops of semen may be released before ejaculation.

As mentioned earlier, spermicidal jellies, creams, foams, and films contain chemicals that kill sperm but are apparently harmless to the woman. They can also be used alone but are only about as effective as withdrawal.

Abstaining from sexual intercourse around the time of ovulation—the **rhythm method**—can help couples reduce the likelihood of pregnancy. If a couple knows the exact time of ovulation, they can time sexual intercourse to prevent pregnancy more precisely.

To practice the rhythm method successfully, couples must first find out when ovulation occurs. Ovulation can be determined by measuring body temperature because in most women body temperature rises slightly after ovulation. Another method used to time ovulation involves taking samples of the cervical mucus. Cervical mucus varies in consistency during the menstrual cycle. By testing its thickness on a daily basis, a woman can tell fairly accurately when she has ovulated.

Because eggs remain viable 12-24 hours after ovulation and sperm may remain alive in the female reproductive tract for up to 3 days, abstinence 4 days before and 4 days after the probable ovulation date should provide a margin of safety (FIGURE 15-11).

Abortion

Some couples may elect to terminate pregnancy through **abortion**, surgical removal of the fetus early in pregnancy.

Abortion is not suitable or morally acceptable to all people. Pro-life advocates argue that abortion

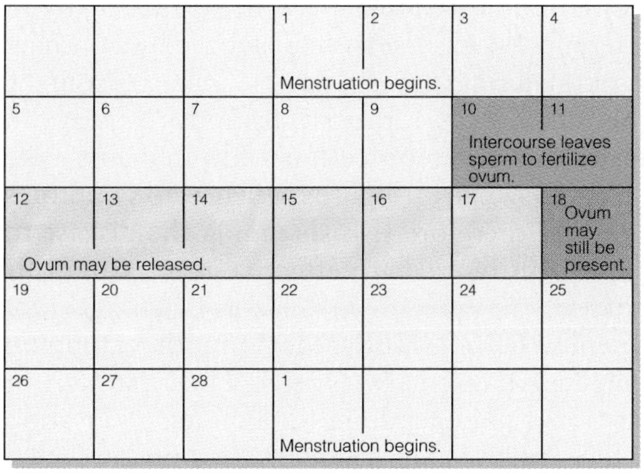

			1 Menstruation begins.	2	3	4
5	6	7	8	9	10 Intercourse leaves sperm to fertilize ovum.	11
12 Ovum may be released.	13	14	15	16	17	18 Ovum may still be present.
19	20	21	22	23	24	25
26	27	28	1 Menstruation begins.			

FIGURE 15-11 The Natural Method The shaded areas indicate an unsafe period for sexual intercourse, assuming ovulation occurs at the midpoint of the cycle.

should be outlawed or severely restricted—that is, allowed only in cases of rape, incest, and threat to the life of the mother. These individuals advise unmarried women to abstain from sexual intercourse or, if they become pregnant, to give birth and either keep the baby or put it up for adoption.

Pro-choice advocates, on the other hand, support abortion. They argue that women should have the freedom to choose whether to terminate a pregnancy or have a child. Abortion, they say, reduces unwanted pregnancies and untold suffering among unwanted infants, especially in poor families, and gives women more options than motherhood. Although pro-choice advocates view abortion as a legitimate means of family planning, they point out that it should not be practiced as a primary means of birth control. Contraception is less costly, less traumatic, and more morally acceptable.

In the first 12 weeks of pregnancy, abortions can be performed surgically in a doctor's office via vacuum aspiration. In this procedure, the cervix is first dilated by a special instrument. Next, the contents of the uterus are drawn out through an aspirator tube.

Vacuum aspiration is a fairly simple and relatively painless procedure. Usually no anesthesia is given. Although women bleed for a week or so after the procedure, they generally experience few complications.

Most abortions are performed by the end of the twelfth week of pregnancy. After 16 weeks, abortions are more difficult and more risky. Solutions of salt or prostaglandins, which stimulate uterine contractions, are injected into the sac of fluid surrounding the fetus to induce labor. The hormone oxytocin may be administered to the woman with the same effect.

Women can also take a pill that prevents fertilized eggs from implanting in the lining of the uterus. It is called RU486, or incorrectly, the abortion pill.

In the United States, there is another chemical treatment that has the same effect. This first pill restricts blood flow to the uterus. Two days after the initial pill is given, a second pill is taken. It prevents the fertilized ovum from implanting.

Sexually Transmitted Diseases

Certain bacteria and viruses can be transmitted by sexual contact. These organisms can penetrate the lining of the reproductive tracts of men and women and proliferate in the moist, warm environment of the body causing **sexually transmitted diseases** (STDs). Most of the infectious agents that cause STDs are spread by vaginal intercourse, but other forms of sexual contact such as anal and oral sex are responsible for their transmission. AIDS, for example, can be transmitted by anal sex as well as vaginal and oral sex. Syphilis is caused by a bacterium that is transmitted similarly.

Although STDs pass from one person to another during sexual contact, the symptoms are not confined to the reproductive tract. In fact, several STDs, including syphilis and AIDS, affect other body systems.

One complicating factor in controlling STDs is that some of them such as gonorrhea may produce no obvious symptoms in an infected individual. As a result, the disease can be transmitted without a person knowing he or she is infected. In others, such as AIDS, symptoms may not appear for weeks after the initial infection. Thus, sexually active individuals can transmit the AIDS virus to many people before they are aware that they are infected.

Gonorrhea (GON-or-REE-ah) is caused by a bacterium that commonly infects the urethra of men and the cervical canal of women. Gonorrhea often causes no symptoms. When they do appear, painful urination and a puslike discharge from the urethra are common complaints in men and women. Symptoms of gonorrhea usually appear about 1–14 days after sexual contact.

These symptoms are indicative of gonnorhea, but as you shall soon see several other sexually transmitted diseases have similar symptoms. To determine the exact cause, one needs to see a doctor who will test to see if the bacterium is present.

Gonorrhea is treated with antibiotics and clears up quickly, usually within 3 to 4 days, if treatment begins early. If left untreated, gonorrhea in men can spread to the prostate gland, making it more difficult to treat. Infections in the urethra lead to the formation of scar tissue. This may narrow the urethra and make urination even more difficult. In some women, bacterial infection spreads to the uterus and uterine tubes, causing the buildup of scar tissue. In the uterine tubes, scar tissue may block the passage of sperm and ova, resulting in infertility. Gonorrheal infections can also spread into the abdominal cavity through the opening of the uterine tubes. If the infection enters the bloodstream in men or women, it can travel throughout the body. Early diagnosis is essential to limit the damage.

Syphilis is an STD caused by a bacterium that penetrates the linings of the oral cavity, vagina, and penile urethra. It may also enter through breaks in the skin. If untreated, syphilis proceeds through three stages. In stage 1, between 1 and 8 weeks after exposure, a small, painless red sore develops, usually in the genital area. Easily visible when on the penis, these sores often go unnoticed when they occur in the vagina or cervix. The sore heals in 1-5 weeks, leaving a tiny scar.

Approximately 6 weeks after the sore heals, individuals complain of fever, headache, and loss of appetite. Lymph nodes in the neck, groin, and armpit swell as the bacteria spreads throughout the body. This is stage 2. It lasts for about 4 to 12 weeks.

As a rule, the symptoms of stage 2 syphilis disappear for several years. Then, without warning, the disease flares up again. This is stage 3. During stage 3, patients experience a loss of their sense of balance and a loss of sensation in their legs. As the disease progresses, patients experience paralysis, senility, and even insanity.

Syphilis can be treated with antibiotics, but only if the treatment begins early. In stage 3, antibiotics are useless. Tissue and organ damage is usually permanent.

One of the most common sexually transmitted diseases, affecting 3 to 10 million people each year, is **chlamydia** (clam-ID-ee-ah). Caused by a bacterium, this disease is characterized by a burning sensation during urination and a discharge from the penis and vagina. If the bacterium spreads, it can cause more severe infection and infertility. Like other STDs, many people experience no symptoms at all and therefore risk spreading the disease to others. Antibiotics are effective in treating this disease.

Genital herpes (HER-peas) is another common STD. It is caused by a virus. The first sign of infection is pain, tenderness, or an itchy sensation on the penis or female external genitalia. These symptoms usually occur 6 days or so after contact with someone infected by the virus. Soon afterward, painful blisters appear on the external genitalia, thighs, buttocks, and cervix, or in the vagina (**FIGURE 15-12**).

The blisters break open and become painful ulcers that last for 1-3 weeks, then disappear. Unfortunately, the herpes virus is a lifelong resident of the body. New outbreaks can occur from time to time, especially when an individual is under stress. Recurrent outbreaks are generally not as severe as the initial one, and, in time, the outbreaks generally cease.

Unlike other STDs, herpes can be transmitted to other individuals during sexual contact only when the blisters are present or just beginning to emerge. When the virus is inactive, sexual intercourse can occur without infecting a partner. Although herpes cannot be cured, physicians can suppress outbreaks with antiviral drugs.

Herpes is not a particularly dangerous STD, except in pregnant women. These women run the risk of transferring the virus to their infants at birth. Because the virus can be fatal to newborns, women are often advised to deliver by cesarean section (an incision made just above the pubic bone) if the virus is active at the time of birth.

Nongonococcal urethritis (YUR-ee-THRIGHT-iss), or NGU for short, is the most common sexually transmitted disease. Moreover, NGU is one of several STDs whose incidence is steadily rising in the United States. Caused by any of several different bacteria, this infection is generally less threatening than gonorrhea, syphilis, and chlamydia, although some infections can result in sterility.

Many men and women often exhibit no symptoms whatsoever and can therefore spread the disease without knowing it. Symptoms resemble those of gonorrhea. NGU can be treated by antibiotics.

The vast majority of Americans carry a virus known as **human papillomavirus** (pap-ILL-oh-mah) or **HPV**. Transmitted by sexual contact, this virus can cause **genital warts**. Genital warts are benign growths that appear on the external genitalia and around the anuses of men and women. Warts also grow inside the vagina of women. Warts generally occur in individuals whose immune systems are suppressed, for example, after long periods of stress.

These warts can remain small or can grow to cover large areas, creating cosmetically unsightly growths. They may cause mild irritation, and certain strains of HPV are associated with cervical cancer in women.

Genital warts can be treated with chemicals or removed surgically—although rates of recurrence are quite high. Getting rid of the virus, however, is impossible, for it resides in the body forever.

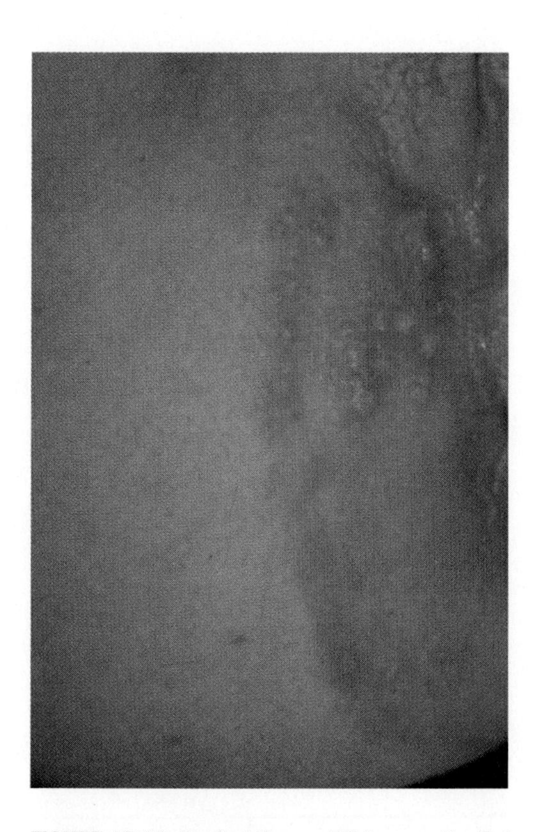

FIGURE 15-12 Genital Herpes Blisters on the External Genitalia and Inner Thigh

Infertility

A surprisingly large percentage (about one in six) of American couples cannot conceive. The inability to conceive (to become pregnant) is called *infertility*. According to some statistics, in about 50% of the couples, infertility results from problems occurring in the woman. Approximately 30% of the cases are due to problems in the man alone, and about 20% are the result of problems in both partners.

If after a year of actively trying to conceive a couple remains unsuccessful, they can consult a fertility specialist who will first check obvious problems such as infrequent sex. Obviously, only intercourse around the time of ovulation is successful. If timing is not the problem, and it rarely is, the physician tests the man's sperm count. A low sperm count is one of the most common causes of male infertility and is easy to test.

A low sperm count may result from overwork, emotional stress, and fatigue. Tobacco and alcohol consumption can also contribute to the problem. The testes are also sensitive to a wide range of chemicals and drugs that reduce sperm production.

If infertility appears to be caused by a low sperm count, a couple may choose to undergo artificial insemination, using sperm from a sperm bank. These sperm are generally acquired from anonymous donors.

If sperm production and ejaculation appear normal, a physician checks the woman, starting with ovulation. If ovulation is not occurring, **fertility drugs** may be administered. Unfortunately, fertility drugs often result in the ovulation of many fertilizable ova, producing 4-6 babies. Most of the multiple births you hear about on the news are the result of fertility drugs.

If ovulation is occurring normally, the physician then examines the uterine tubes to see if they are obstructed. In some instances, a previous gonorrheal or chlamydial infection that spread into the tubes may have caused scarring that obstructed the passageway. In such instances, couples may be advised to adopt a child or to try *in vitro* (in VEE-trow) fertilization.

During ***in vitro*** **fertilization**, eggs are removed from the woman and fertilized by the partner's sperm outside her body. The fertilized ovum is then implanted in the uterus of the woman. Besides being expensive and time-consuming, this procedure has a low success rate.

Female reproduction, like male reproduction, is vital to the continuation of our species. But as you will see in the next two modules we humans have been quite successful in continuing our kind. Too many people can actually have a harmful impact on the long-term future of humankind.

Principles of Ecology

Most of us live our lives in cities and towns seemingly apart from nature. Despite this, our lives depend heavily on nature for just about everything: from our morning coffee to the clothes we wear to the wood that's used to build our homes. Resources from nature are vital to our economy, too.

Nature also provides many free services that very few people ever think about. Plants, for instance, replenish the oxygen we breathe and protect the land near our homes from soil erosion. Swamps along streams help purify the water in streams that we swim in and drink from. Birds fluttering around our gardens help to control insect populations.

Clearly, nature serves us well. Because of this, damage to the environment can be quite costly.

In this section, which discusses principles of ecology, you'll learn how natural systems work and why they are so important to us.

Ecology

Ecology is a branch of science. Its main focus is on how organisms—plants, animals, and microbes—interact with one another and how they interact with their physical and chemical environment. Ecology also looks at the ways organisms affect their environment for better and for worse.

The Biosphere

The science of ecology focuses on systems. The largest biological system on Earth is called the **biosphere** (BUY-oh-sfear). The biosphere can be thought of as the thin skin of life on the planet.

The biosphere extends from the bottom of the ocean to the tops of the highest mountains. Although that may seem like a long way, it's not—at least, when compared to the size of the Earth. In fact, if the Earth could be reduced to the size of an apple, the biosphere would only be about the thickness of its skin. Life exists throughout the biosphere, but is rare at the extremes where conditions for survival are difficult.

Virtually all life in the biosphere is powered by sunlight. And all of the matter in the biosphere, such as the carbon atoms in the proteins in the muscles of your arm, are recycled over and over again. Long after you're gone, those atoms will still be in service, forming the body of another living thing.

The biosphere stretches over huge land masses, the continents, and huge bodies of water, the oceans primarily. But the biosphere is not uniform throughout. In fact, if you could travel in the space shuttle, looking down on planet Earth, one of the things you would notice is that the land masses are actually subdivided into very distinct areas, such as the deserts and rain forests. Each of these large and ecologically distinct regions is known as a **biome**. Each biome has a distinct climate and supports a unique set of plants and animals. **FIGURE 16-1** shows the biomes of North America, including deserts, grasslands, and heavily forested areas like the eastern deciduous forest biome.

Another thing you would notice as you soar above the Earth is that the majority of the planet is covered with water. Just like the continents, though, the oceans consist of ecologically distinct zones, known as **aquatic life zones**, such as the deep ocean and coral reefs. Conditions in each aquatic life zone differ from the others, and each aquatic life zone supports a unique group of plants and animals adapted to the conditions within its zone.

Humans inhabit all biomes on Earth, and benefit in numerous ways from these regions as well as the aquatic life zones, as mentioned in the introduction of this module. Forest biomes, for instance, provide us with wood and raw materials to make paper and, lest we forget, places to camp and hike and listen to the sounds of birds. Forests also replenish oxygen in the air we breathe. The aquatic life zones of the ocean provide fish that feed millions of people the world over and many other species that depend on them.

Ecosystems

The biosphere is a large and complex system. We call such systems **ecosystems**, which is short for **ecological systems**. Ecosystems consist of many living organisms: plants, animals, and microorganisms. But they also consist of many nonliving components, such as sunlight, rainfall, and temperature.

As you will see in the next module, humans often alter the living and nonliving components of ecosystems, making it difficult or impossible for organisms to survive in them. If conditions change, some species may decline in number. If changes are drastic, some species may perish altogether or be forced to seek other suitable living space. Even humans can be driven from an area by severe changes in local ecosystems.

Although plants and animals are sensitive to all of the nonliving components of their ecosystems, one factor often turns out to be more important than others in regulating the growth of an entire ecosystem. This factor is called a **limiting factor**. In freshwater lakes and rivers, dissolved phosphate is the limiting factor. Phosphate is required by plants and algae for growth, but phosphate concentrations are naturally low. As a result, plant and algal growth is normally held in check. When phosphate is added to a body of water, for example, from human waste and detergents released by a sewage treatment plant, plants and algae proliferate. Algae often form dense surface mats, blocking sunlight. As a result, plants rooted on the bottom of the lake die. When they die, they are no longer able to produce oxygen. When oxygen levels in the water decline, fish and other species may die.

On land, precipitation (rain and snow) tends to be the limiting factor. At any given temperature, the more moisture that falls on the land, the richer the plant and animal life.

Organisms frequently live in groups within specific areas. These groups are called **populations**. The members of a population may live together throughout all or much of the year, like deer, or may live separate lives over a large area, keeping to themselves except for mating as in the case of black bears in North America. In any given ecosystem, several populations exist together and form a **biological community**.

FIGURE 16-1 The Biomes

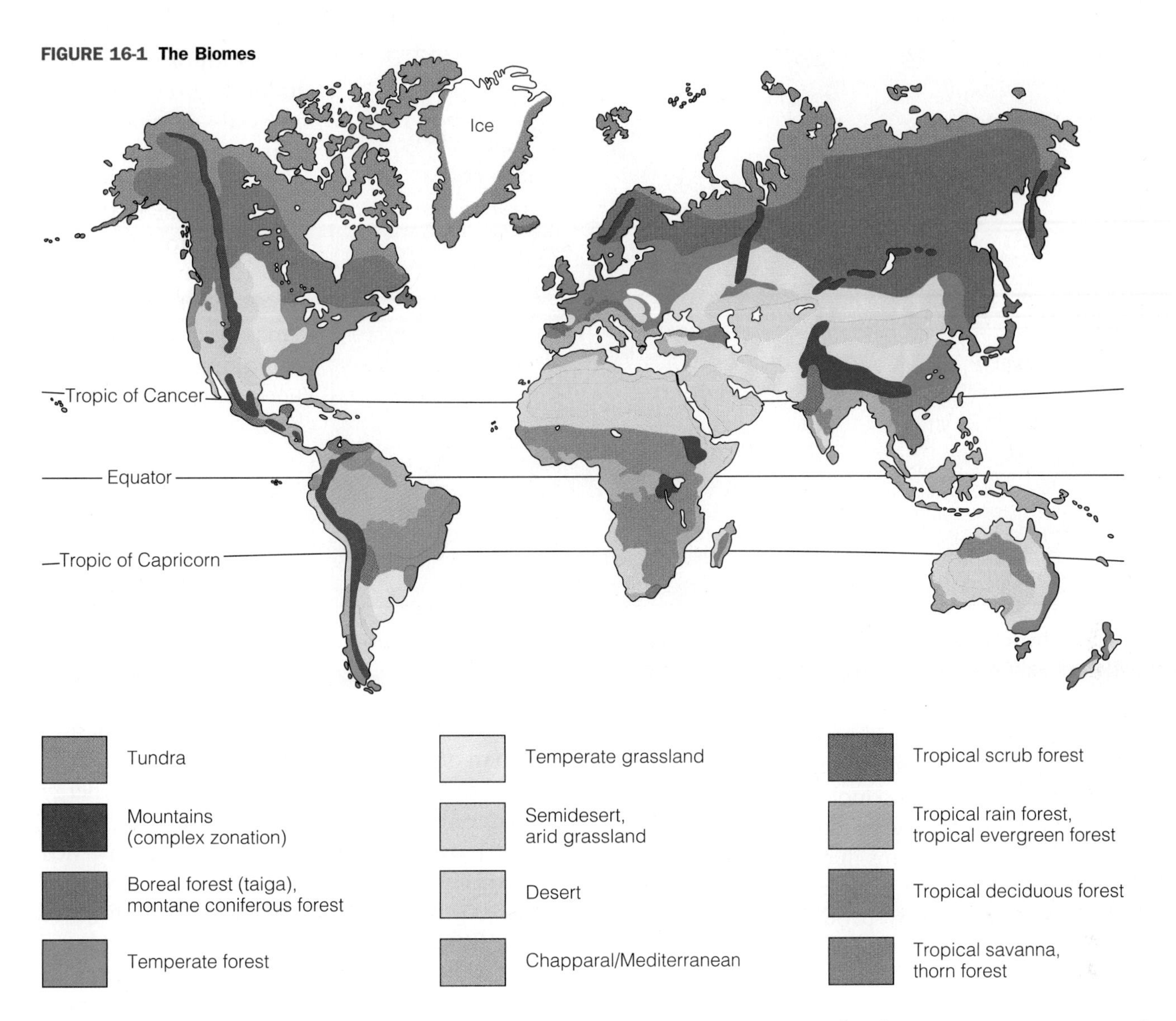

Competition in Ecosystems

The place in which an organism lives is called its **habitat.** But within its habitat each organism interacts with others, sometimes competing with members of its own species or with members of other species for food, water, and other resources. Competition can be intense or mild, depending on the abundance of resources and the size of the populations.

Humans also compete with many other species for food and habitat. However, humans have an advantage over most species, in large part, because of our advanced technology. For example, commercial fishing operations compete with many forms of sea life, such as sea lions and seals, for fish. Such activities not only threaten those species we compete with but can, if not carefully managed, threaten our own food sources. As an example, overfishing of the world's oceans has already destroyed dozens of the fish populations (**FIGURE 16-2**). In addition, continued pressure on fish population means that many commercial fish species are currently on the decline and could be depleted if measures are not taken—and soon—to reduce fishing pressure.

Saving other species is more than a humanitarian act, then. It is an act of self-protection—it helps protect valuable resources we and future generations depend on.

How Ecosystems Work

Virtually all life in ecosystems depends on a group of organisms known as *producers*. **Producers** are

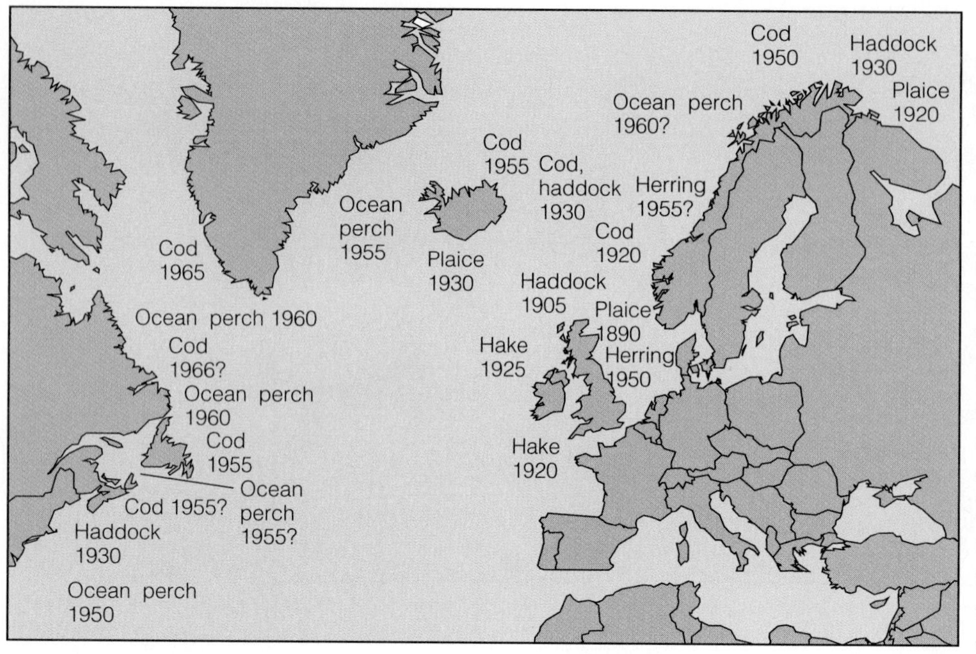

FIGURE 16-2 Depletion of North Atlantic Fisheries The dates on the map indicate the approximate time when various commercial fisheries were depleted by overfishing.

organisms such as algae and plants that absorb sunlight and use its energy to make food molecules using carbon dioxide and water. This process is called **photosynthesis.** Food molecules made during photosynthesis nourish the producers, but also feed all the rest of the organisms in ecosystems.

One of the groups of organisms that lives off the bounty generated by producers is the **consumers**. You and I fit into this group, as do all other animals.

Ecologists place consumers into four general categories, depending on the type of food they eat. Deer, elk, and cattle feed solely on plants and are called **herbivores**. Wolves and bears feed on herbivores and other animals and are known as **carnivores.** Humans and a great many other animal species that subsist on a mixed diet of plants and animals and are known as **omnivores**. The final group feeds on animal waste or the remains of plants and animals. These organisms are called **decomposers.** This important group includes many bacteria, fungi, and insects.

Organisms that feed on one another form **food chains** (**FIGURE 16-3**). Ecosystems typically contain two types of food chains: grazer and decomposer.

Grazer food chains begin with plants and algae, the producers. These organisms are consumed by herbivores, or grazers; hence the name. In grazer food chains, herbivores are typically eaten by carnivores or omnivores.

Decomposer food chains begin with dead material—either animal wastes (feces) or the remains of plants and animals (**FIGURE 16-4**).

As just noted, grazer and decomposer food chains function together to make ecosystems work. For example, waste products from the grazer food chain enter the decomposer food chain. Nutrients liberated by the decomposer food chain in turn enter the soil and water and are taken up by plants in the grazer food chain.

Because there are many food chains in an ecosystem and because organisms tend to belong to more than one food chain, food chains are usually woven into larger **food webs** (**FIGURE 16-5**). Energy and materials flow through food webs.

In the biosphere, almost all of the energy needed by organisms comes from the sun. Solar energy is captured by plants and algae and used to produce organic food molecules. Energy from the Sun is stored in these molecules in chemical bonds so when a deer nibbles on grass, it is not only obtaining the food it needs, it is acquiring energy. The energy and food molecules in the deer are then passed on to coyotes, wolves, and bear that feed on deer.

Energy acquired by plants and animals is used to run many cellular processes, including muscle contraction, that makes movement possible. But eventually all energy consumed by organisms is converted to heat that then escapes into outer space. As a result, energy is said to flow in one direction through food webs. It cannot be recycled.

In sharp contrast, nutrients that are taken up by plants and algae move through food webs, from one organism to another but are continually recycled. Carbon atoms from carbon dioxide, for example, are taken up by plants to make food molecules that are passed on from one organism to the next in a food web. When an animal dies, the carbon atoms are released as the body decays, thus

FIGURE 16-3 Simplified Food Chains An aquatic and terrestrial food chain are shown here.

re-entering the environment. Nutrients like carbon atoms may also be recycled from waste, urine and feces, deposited by animals.

One way or another, all nutrients eventually make their way back to the environment for recycling. Each new generation of organisms therefore relies on recycled material from previous generations. The carbon atoms in your body, for instance, have been recycled many times since the beginning of life on Earth. Who knows, some of those atoms may have been in the very first cell that lived in the shallow seas!

Ecologists classify organisms in a food chain according to their position, or **trophic level** (TROE-fic; literally, "feeding" level). The producers comprise the base of the grazer food chain and are members of the first trophic level. The grazers are members of the second trophic level. Carnivores that feed on grazers belong to the third trophic level.

Most terrestrial food chains are limited to three or four trophic levels. Longer terrestrial food chains are quite rare. The reason for this is that food chains generally do not have a large enough producer base to support many levels of consumers. Therefore, there's not enough food being created by producers to support a

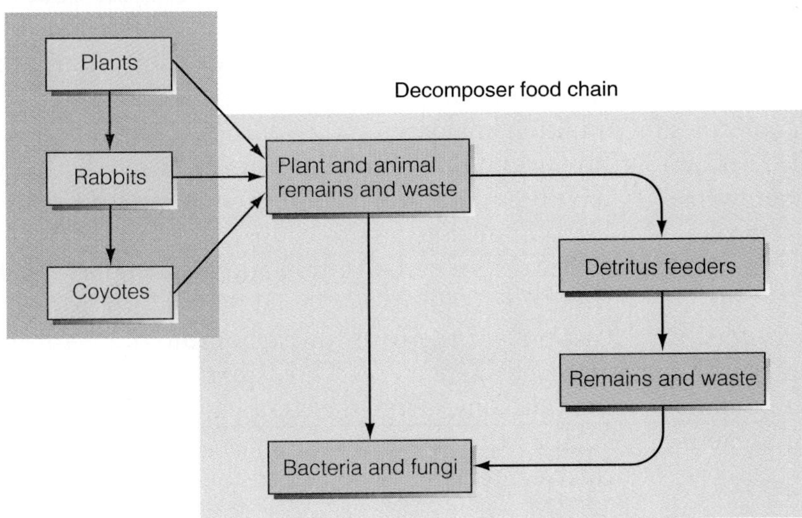

FIGURE 16-4 Food Chains A grazer food chain and a decomposer food chain, showing the connection between the two.

FIGURE 16-5 A Food Web

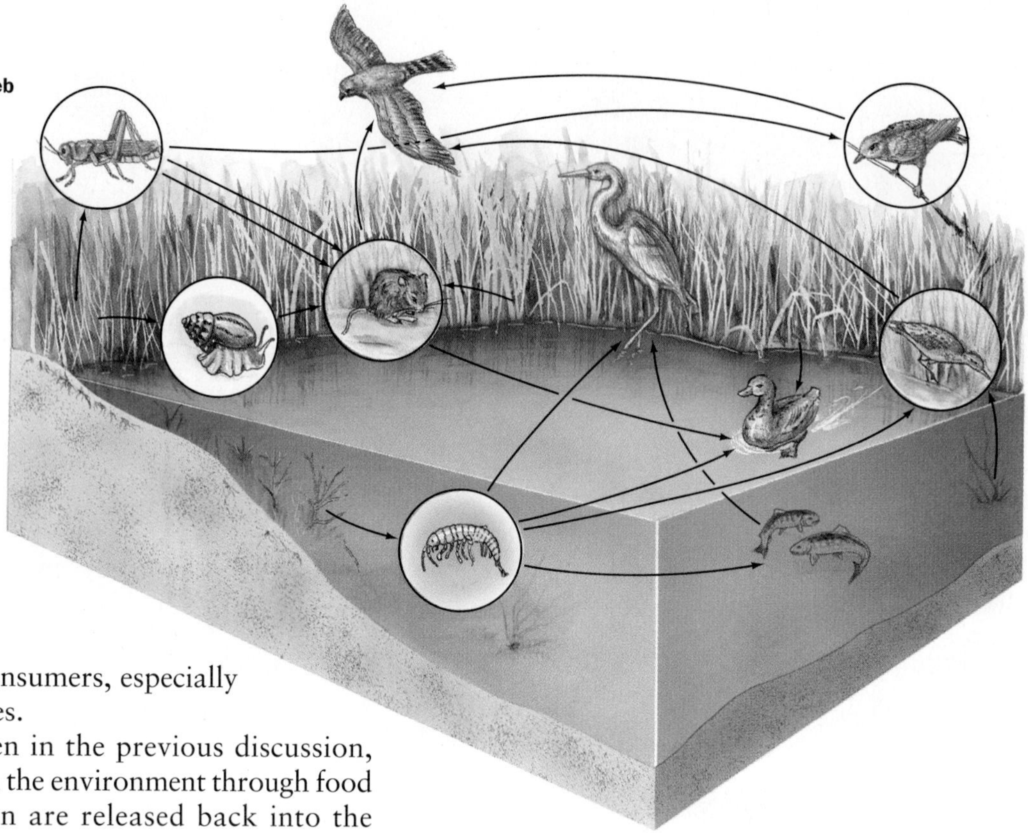

large number of consumers, especially the higher-level ones.

As you have seen in the previous discussion, nutrients flow from the environment through food webs, and are then are released back into the environment. This circular flow constitutes a **nutrient cycle**. Dozens of global nutrient cycles operate continuously to ensure the availability of chemicals vital to all living things, present and future. Unfortunately, however, a great many human activities disrupt them. These activities, in turn, may profoundly influence the survival of many other species. To understand how, let's take a look at two of the most important nutrient cycles, the carbon and nitrogen cycles. Remember, however, that these are only two of several dozen nutrient cycles essential to life.

The Carbon Cycle. The **carbon cycle** illustrated in FIGURE 16-6 begins with free carbon dioxide found in the air we breathe and the water in oceans, lakes, and rivers. Plants growing on land and in waters absorb carbon dioxide and water. They then use energy from sunlight to produce food molecules from these raw materials. The food molecules are passed from one organism to another in food webs. Along the way, some food molecules are broken down to produce energy. Others are used to build body parts that are broken down when organisms die and decompose. Others are lost in waste. During the breakdown of these molecules, carbon dioxide is released. It then reenters the air or water where it can be used again by plants or algae to make more food.

For much of the Earth's history, carbon dioxide levels remained fairly constant in the atmosphere because production from wastes and other sources equaled the amount being consumed. Over the past 100 years, carbon dioxide production from human sources has increased steadily, causing atmospheric concentrations to increase. Humans produce large quantities of carbon dioxide during the combustion of fossil fuels, such as home heating oil, jet fuel, and gasoline.

Deforestation—cutting trees down without replacing them—also increases carbon dioxide levels in the environment. Why? Trees absorb enormous amounts of carbon dioxide during photosynthesis. When they're cut down and not replaced, carbon dioxide levels increase. In addition, many forests are burned to clear the land for farming or human settlement after the trees have been cut, further adding to the carbon dioxide levels in the Earth's atmosphere.

Because of deforestation and the combustion of fossil fuels, humans add over 7 billion tons of carbon—in the form of carbon dioxide—to the atmosphere each year. Three quarters of the carbon dioxide comes from the combustion of fossil fuels such as the gasoline in our cars; the remaining quarter stems from deforestation.

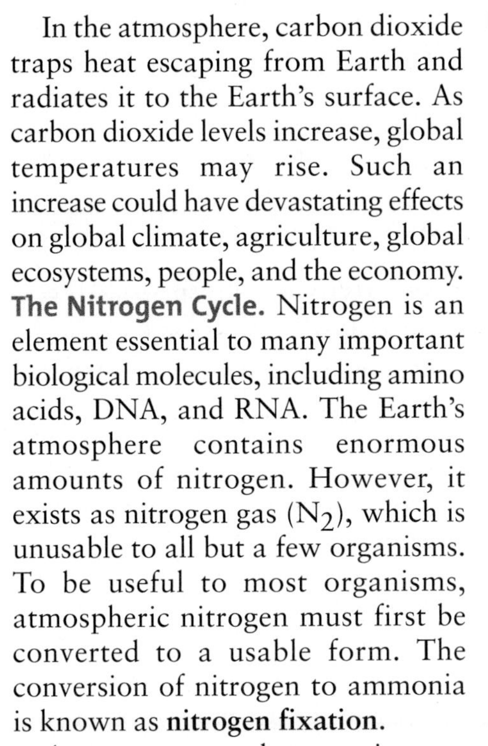

In the atmosphere, carbon dioxide traps heat escaping from Earth and radiates it to the Earth's surface. As carbon dioxide levels increase, global temperatures may rise. Such an increase could have devastating effects on global climate, agriculture, global ecosystems, people, and the economy.

The Nitrogen Cycle. Nitrogen is an element essential to many important biological molecules, including amino acids, DNA, and RNA. The Earth's atmosphere contains enormous amounts of nitrogen. However, it exists as nitrogen gas (N_2), which is unusable to all but a few organisms. To be useful to most organisms, atmospheric nitrogen must first be converted to a usable form. The conversion of nitrogen to ammonia is known as **nitrogen fixation.**

As **FIGURE 16-7** shows, nitrogen fixation occurs, in part, in the roots of certain plants, such as peas, beans, clover, and alfalfa. These plants, known as **legumes**, have small swellings on their roots called **root nodules.** Inside the nodules live certain bacteria that convert atmospheric nitrogen to ammonia. Ammonia is also produced by certain types of bacteria (known as cyanobacteria) in the soil.

FIGURE 16-6 The Carbon Cycle

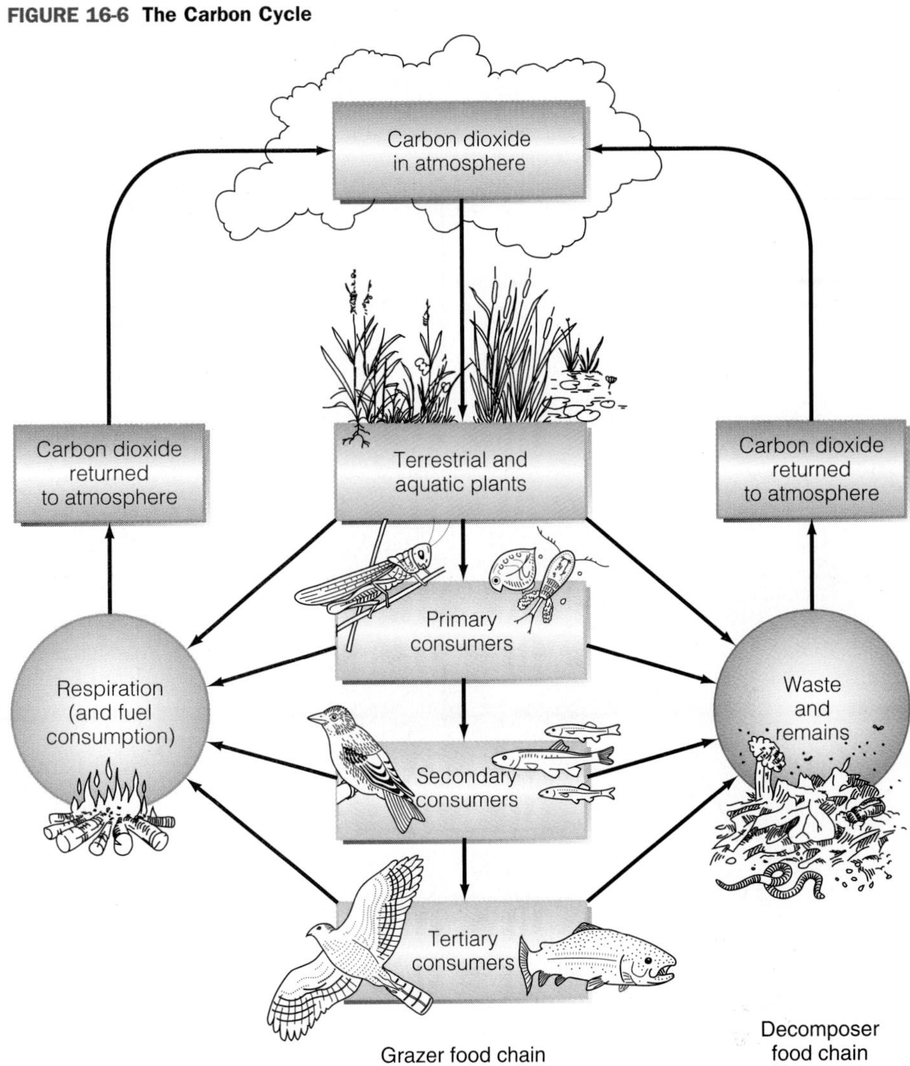

Ammonia is then converted to another chemical, called nitrite, which is then converted to nitrate. Nitrates are incorporated by plants and used to make amino acids and nucleic acids.

Nitrate in soil is not only made from nitrogen in the atmosphere, it comes from decaying plants and animals as well as the waste of animals. Ammonia from these sources is converted to nitrite, then to nitrate and reused.

Humans alter the nitrogen cycle in several ways. For example, some farmers apply too much fertilizer to their land. Fertilizer contains nitrogen. During heavy rains, the excess nitrogen may be washed into lakes and streams. Sewage from cities and towns and waste from livestock operations also contain lots of nitrogen that may be released into waterways from sewage treatment plants.

In waterways, nitrogen stimulates the growth of algae and aquatic plants. As a result, rivers and lakes may become congested with dense mats of vegetation, making them unnavigable. Sunlight penetration to deeper levels is impaired by the growth of algae that form dense mats on the surface. This causes oxygen levels in deeper waters to decline. In the autumn, when the algae and aquatic plants die, they decay. This process uses oxygen, and thus causes dissolved oxygen levels to fall further, threatening aquatic animal life.

Ecosystem Homeostasis

In earlier modules, I pointed out that human health depends on internal balance. A breakdown of the processes that contribute to this balance can lead to disease and, if severe enough, death. Ecosystems

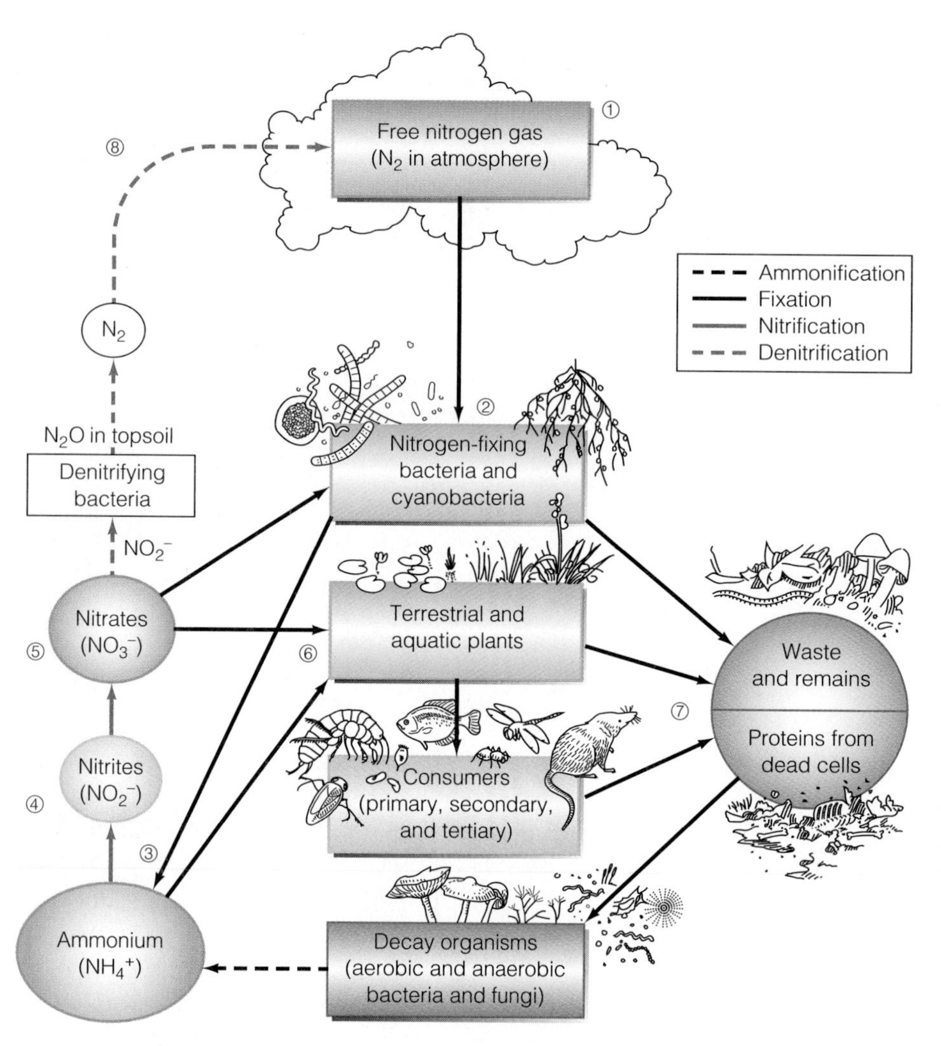

FIGURE 16-7 The Nitrogen Cycle Nitrogen in the atmosphere is converted to NH_3 (ammonia) by bacteria and cyanobacteria in soil. Ammonia is converted to nitrates and taken up by plants. To follow the flow of nitrogen, follow the circled numbers.

also depend on mechanisms such as the carbon and nitrogen cycles that help maintain balance.

Ecosystem balance is not a steady state. In other words, nature does not remain the same from one day to the next. In ecosystems, populations grow and decline in natural cycles but remain more or less the same over the long term.

Ecosystem balance results from many forces that act on individual populations within an ecosystem (**FIGURE 16-8**). Some factors tend to cause populations to grow. Favorable weather and ample food supplies are two such factors.

The second set of factors are those that cause populations to decline. Adverse weather, lack of food, disease, and predation, for instance, can cause populations to decline.

Numerous growth and reduction factors operate at the same time in ecosystems. In the midwestern United States, for example, grasses grow well during wet years. Because of the abundance of food, mouse populations expand. This increase, in turn, results in larger populations of hawks, which feed on rodents. The more food that is available, the more hawk offspring will survive.

In this simplified example, favorable weather conditions shift ecosystem balance. Conditions are partly restored, however, by the increase in the hawk populations. As the mouse population falls, the number of predators decreases, restoring balance.

Changes such as these occur with great regularity in ecosystems. If changes are minor, they are of little consequence. Ecosystems can recover easily.

Natural Succession. The ability of ecosystems to recover from minor changes results in a measure of stability. However, some changes can drastically upset ecosystem balance. Natural changes such as volcanic eruptions and forest fires and human activities, such as clear cutting or farming, can dramatically affect ecosystems. Although ecosystems can recover from such events, the process can be quite slow. A forest that is cleared, then planted with crops, for example, may recover if farming ceases, but the process will require 70 years or more. This process of restoration is called **natural succession.**

Succession, in general, is a process of change in which one biological community is replaced by another until a mature ecosystem is formed. Two types of succession exist: primary and secondary.

Primary succession occurs where no biotic community previously existed—for example, when deep-sea volcano erupts to form a barren island. Over tens of thousands of years, rich plant and animal community can form on the lifeless volcanic rock. How does life form?

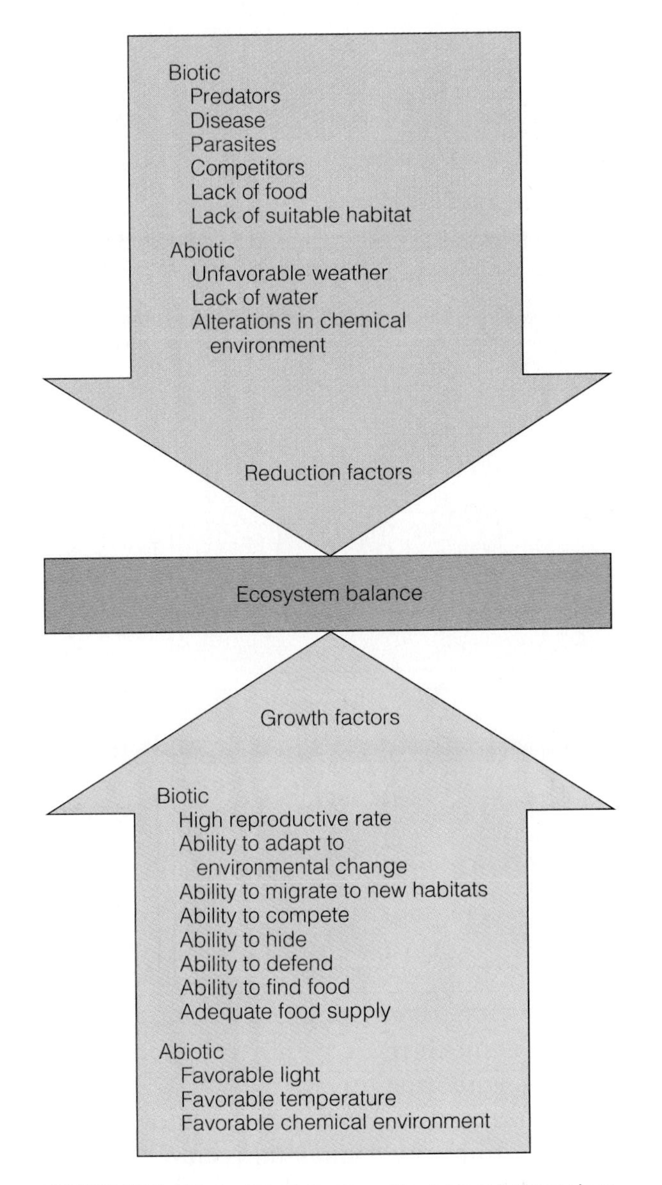

FIGURE 16-8 Ecosystem Balance Ecosystem balance is a complex equilibrium brought about by biotic and abiotic growth and reduction factors.

Plants come from seeds that are carried to the islands by waves or dropped in the feces of birds flying overhead. Over time, the original plants spread to cover the entire island. New plant species evolve from these plants. Birds may settle on the island as well. In the ensuing years, new bird species may evolve from them.

Secondary succession occurs when a biological community that had previously existed is destroyed by natural forces or human actions and then regrows. **FIGURE 16-9** shows secondary succession in an abandoned eastern farm field. As illustrated, hardy crabgrass first invades the abandoned field. Unlike many other plants, it can tolerate hot, sunny land.

In the next twenty years, other plants begin to grow from seeds from several different sources. Some may already be in the soil. Others are blown in from neighboring fields or carried in by animals.

Over the next thirty years, pine trees sprout and begin to grow. As the pines become larger, they begin to create shade. Next, shade-tolerant hardwood trees begin to grow. These trees eventually grow so large that they "shade out" many of the pine trees.

Seventy years after this process begins, a mature forest is formed, consisting mostly of hardwoods and an occasional pine—as it did before the land was cleared for farming.

Secondary succession occurs much more rapidly than primary succession because soil is already present. In primary succession, soil must be formed from rock, gravel, or sand, and this can take 100-1000 years.

Just as damage to the human body cannot always be repaired, damage to ecosystems, if severe enough, may not be remedied by secondary succession. In Vietnam, for example, thousands of acres of tropical forests were destroyed by allied forces. They had sprayed Agent Orange, an herbicide, on them from planes and helicopters during the Vietnam war to reduce the risk of ambush. Broadleaf trees died quickly. After the war, the forests were invaded by a hardy brush that some scientists think may prevent mature forests from regrowing.

Species Diversity

As ecologists study ecosystems, one of the factors they look at is species diversity. **Species diversity** is a measure of the number of different species in a given ecosystem. The more species there are, the more diverse an ecosystem is.

Species diversity varies from one biome to the next. Generally those with the most favorable living conditions, for example, the warmest temperatures, tend to have the greatest diversity. Tropical rainforests, which lie along the warm equator, for example, have the highest species diversity in the world.

Unfortunately, many human activities decrease species diversity of ecosystems. Cutting tropical rainforests, for instance, and replanting them with

FIGURE 16-9 Secondary Succession On an abandoned eastern farm (top left), nature begins the process of restoring a biotic community.

one type of tree, for example, bananas, results in a drastic decline in species diversity. Simplifying an ecosystem (removing species) often tends to make it unstable. A banana plantation, for example, is much more prone to insect damage than the neighboring forest.

In the next section, we'll put this knowledge to use as we study environmental issues. You'll see how an understanding of ecology can be helpful in managing ecosystems—and ourselves—to create a sustainable future.

Environmental Issues: Population, Pollution, and Resources

Environmental problems facing us are commonly referred to as the "environmental crisis." To some people, the phrase "environmental crisis" seems like an exaggeration. But many experts who have studied the trends believe that the problems are severe enough to warrant the term "crisis."

This module describes the world's leading environmental problems. It should help you make up your mind about the appropriateness of the term "environmental crisis." It also discusses solutions that could help us create a sustainable future.

Overshooting the Earth's Carrying Capacity

Environmental problems occur in both the rich, industrialized nations and the poor, less-developed nations. Although the problems vary from one nation to another, they all are signs of a common root cause: human society is exceeding the Earth's carrying capacity.

Carrying capacity is the number of organisms an ecosystem can sustain or support indefinitely. It is determined by (1) food production, (2) resource supply, and (3) the environment's ability to assimilate pollution. As you study the environmental problems described in this module, you will see many signs that we are exceeding the carrying capacity of ecosystems, locally, regionally, even globally.

Overpopulation: Problems and Solutions

One reason we're exceeding the carrying capacity, say scientists, is that we are overpopulated. **Overpopulation** is defined as a condition in which there are too many members of a population to live sustainably—or within the carrying capacity.

Today, the world's population is about 6.5 billion and is increasing at a rate of 1.4% per year. Although the growth rate may seem small, it translates into 87 million new people every year or about 240,000 people being added to the world population every day. If

FIGURE 17-1 Poverty Abounds About three-fifths of the world's people live in poverty. Nearly one of every five people on this planet does not have enough to eat. Many live in makeshift shelters like these in Rio de Janeiro.

the current rate continues, world population is expected to reach 9 billion people by the year 2050—well within most reader's lifetime.

Overpopulation is a problem in virtually all countries, rich and poor. Today, the most rapid growth is occurring on three continents: Africa, Asia, and Latin America. In Europe, the population is actually shrinking. In North America, it is growing, albeit slower than in Africa, Asia, and Latin America.

Most people view overpopulation as a problem of the less developed nations. One of the problems it creates is food shortage. As the global human population grows, many nations are finding it more and more difficult to meet rising demands for food. Starvation abounds in the less developed nations where an estimated 12 million people, many of them children, die each year from malnutrition and starvation or from diseases worsened by hunger.

In many less developed countries, large segments of the population are living in extreme poverty. People live in makeshift homes or on the streets (**FIGURE 17-1**). About 2 billion people are on the edge of poverty, with barely enough to eat.

Overpopulation is also considered by many to be as serious in the United States as it is in poor countries such as Bangladesh. The reason for this is the fairly high standard of living in this rich, industrialized country. As a result, each American uses 20-40 times as many resources as an individual in Bangladesh. Overpopulation combined with our enormous resource demand result in a host of problems discussed in this chapter from species extinction to resource depletion to massive pollution of the air and water.

Further growth in all countries will very likely worsen problems. Unfortunately, little is being done to stem the swelling tide of humanity.

Exponential growth. The human population has not always been so large, nor has it always grown so rapidly. As **FIGURE 17-2A** shows, it was not until the last 200 years that global human population began to skyrocket. This sudden increase was the result of better sanitation, improvements in medicine, and advances in technology.

The graph of human population growth in **FIGURE 17-2** is J-shaped and is known as an **exponential curve**. What is exponential growth?

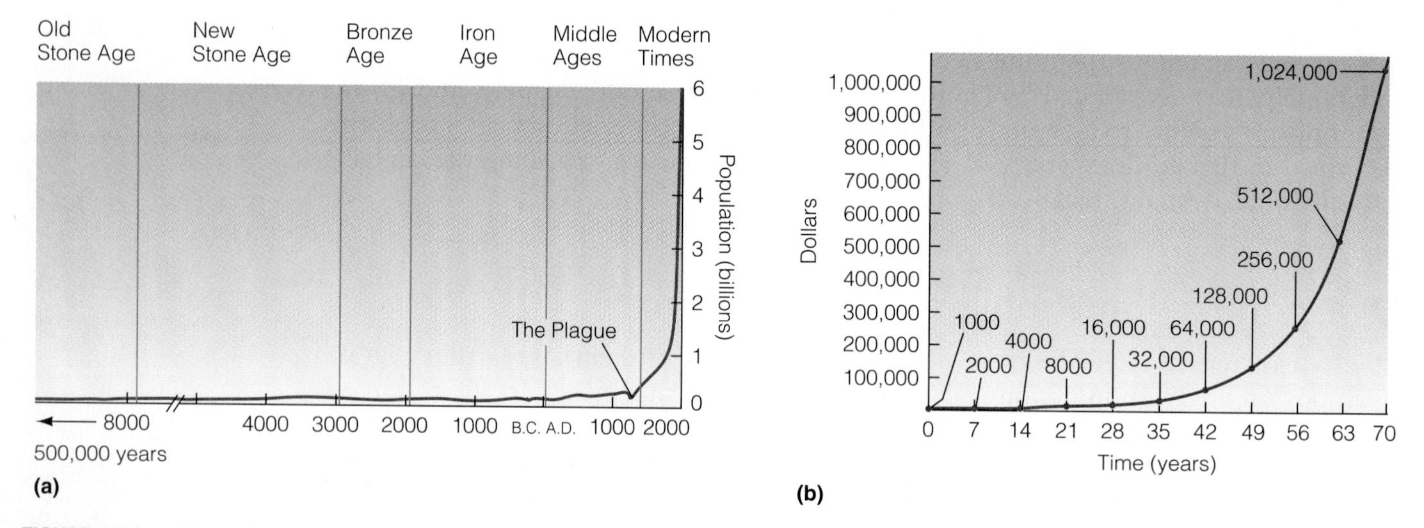

FIGURE 17-2 Exponential Growth of the World Population *(a)* World population. *(b)* Exponential growth of a bank account starting with $1000 at 10% interest.

Let's begin by considering an example. Suppose your parents invested $1000 in a savings account at 10% interest the day you were born. Suppose also that the interest your money earned was applied to the balance, so that it too earned interest. If this were the case, your bank account would be said to be growing exponentially. **Exponential growth** occurs anytime a value, such as population size, grows by a fixed percentage. But only if the interest (annual increase) is applied to the base amount.

Exponential growth is deceptive. In this example, the growth in your account seems small at first. At 10% interest, the $1000 will double in 7 years, yielding $2000 (FIGURE 17-2B). It will double again in 7 years, yielding $4000. The next doubling, occurring when you are 21 years old, will give you $8000. By age 42, the account will have grown to $64,000. By the time you were 49, your account would hold $128,000. At age 56, you'd have over $250,000. If you waited 7 more years, the account would grow to over $500,000. In 7 more years, at 70 years of age, you'd be a millionaire.

What is deceptive about exponential growth is that it begins slowly, then seems to go wild, even though the rate of growth is constant throughout the entire period. In this example, it took 49 years for your account to grow from $1000 to $128,000. In the next 21 years, however, your account grew by $900,000. Why?

Although your account doubled every 7 years, it was not until the base amount was very large that the doublings amounted to much. After the base amount reached this point, each doubling yielded what appeared to be incredible gains.

The human population has taken over 3 million years to reach its current size of about 6.5 billion, but because of exponential growth, it could increase by another 6.5 billion in the next 50 years, if current growth continues. (It is not expected to.)

Pollution and resource demand tend to increase in sync with human population growth, creating concern among scientists and world leaders. Even though supplies of many resources seem large, exponential growth in demand could deplete them in a very short time. For example, a resource with a billion-year life span at the current rate of consumption would be gone in 500 years if the rate of demand increases 5% per year.

Solving World Hunger. Solving world hunger, one of the most visible signs of overpopulation, will require many actions (TABLE 17-1). One of the most important steps is to reduce the rate at which our populations are growing. This could help less developed countries meet the demand for food but could also improve living conditions. In industrial nations, reducing the rate of growth could result in reductions in the demand for vanishing resources and could help reduce pollution.

Just slowing the growth of human population, say many observers, will not be sufficient. To create a sustainable future, humankind may also need to reduce the size of the human population. This, they say, will permit human society to live within the Earth's carrying capacity.

The principal way to reduce population size is through attrition, that is, reducing the number of births through humane measures. If the number of births each year falls below the number of deaths, populations will begin to shrink.

Reducing the population size of a country, indeed the world, is not as difficult as many might believe. Already, approximately 50 countries in Europe have reached a stage of extremely slow growth, no growth, and even shrinkage, among them Hungary, Denmark, Italy, and Austria. The decline in population growth is achieved, in large part, because couples are having smaller families. Globally, birth rates also appear to be falling. And in some countries, adverse environmental conditions, disease such as AIDS, and war are causing death rates to rise, further decreasing population size.

TABLE 17-1

Some Solutions to Alleviate World Hunger
Reduce population growth.
Reduce soil erosion.
Reduce desertification.
Reduce farmland conversion.
Improve yield through better crop strains.
Improve yield through better soil management.
Improve yield through fertilization.
Improve yield through better pest control.
Reduce spoilage and pest damage after harvest.
Use native animals for meat production.
Tap farmland reserves available in some countries.

Resource Depletion: Eroding the Prospects of All Organisms

Creating a sustainable future also relies on measures to protect renewable resources such as forests, farmland soils, water, and oil.

Forests. At one time, tropical rain forests covered a region about the size of the United States (FIGURE 17-3). Today, about half of those forests are gone. Because of an exponential increase in population size and the resulting increase in timber harvesting to meet human needs, most of the remaining tropical rain forests could be lost within your lifetime. Along with them, thousands of species, perhaps as many as a million, could vanish.

Forests are also declining in many other areas, outside of the tropics. Because global tree-cutting exceeds reforestation and because population and demand for resources continue to climb, many scientists are concerned for the fate of the world's forests.

Today, many environmentalists are urging better management of forests, including tree planting, and measures to reduce our demand for wood and wood products, such as building smaller homes and paper recycling. Efforts such as these will help ensure a supply of wood and wood products for future generations but also protect the habitat of wild species. And, they help protect the vital services forests provide to human society, such as purifying the air, reducing carbon dioxide buildup in the atmosphere, protecting watersheds, and reducing soil erosion.

Agricultural Soils. Soil erosion is a natural occurrence, but it is often accelerated by human activities such as farming, construction, and mining (FIGURE 17-4). In the past 100 years, about one-third of the topsoil on American farmland has been eroded by water and wind. Soil erosion continues today, but at a slower rate in the United States thanks to conservation measures by farmers. However, soil erosion still exceeds replacement levels in some areas. Soil erosion is even more serious in many less developed countries.

Soil erosion decreases the fertility of farmland and may make some land useless for farming. Farmland is also lost to development: highways, new suburbs, shopping malls, and so on (FIGURE 17-5). In the United States, an estimated 3500 acres of rural land actual or potential farmland, pasture,

FIGURE 17-3 Tropical Rain Forests *(a)* This map shows the steady decline in tropical forests and projected remaining trees in the year 2000. *(b)* Deforestation: a desperate effort to stave off poverty results in an environmental disaster. These farmers live near the Andasibe reserve in Madagascar, an island where the per capita income is less than $250 a year. Clearing the rain forest allows these impoverished farmers to plant rice and graze cattle. But the environmental price may be too steep. In Madagascar alone, 80% of the rain forest has been destroyed.

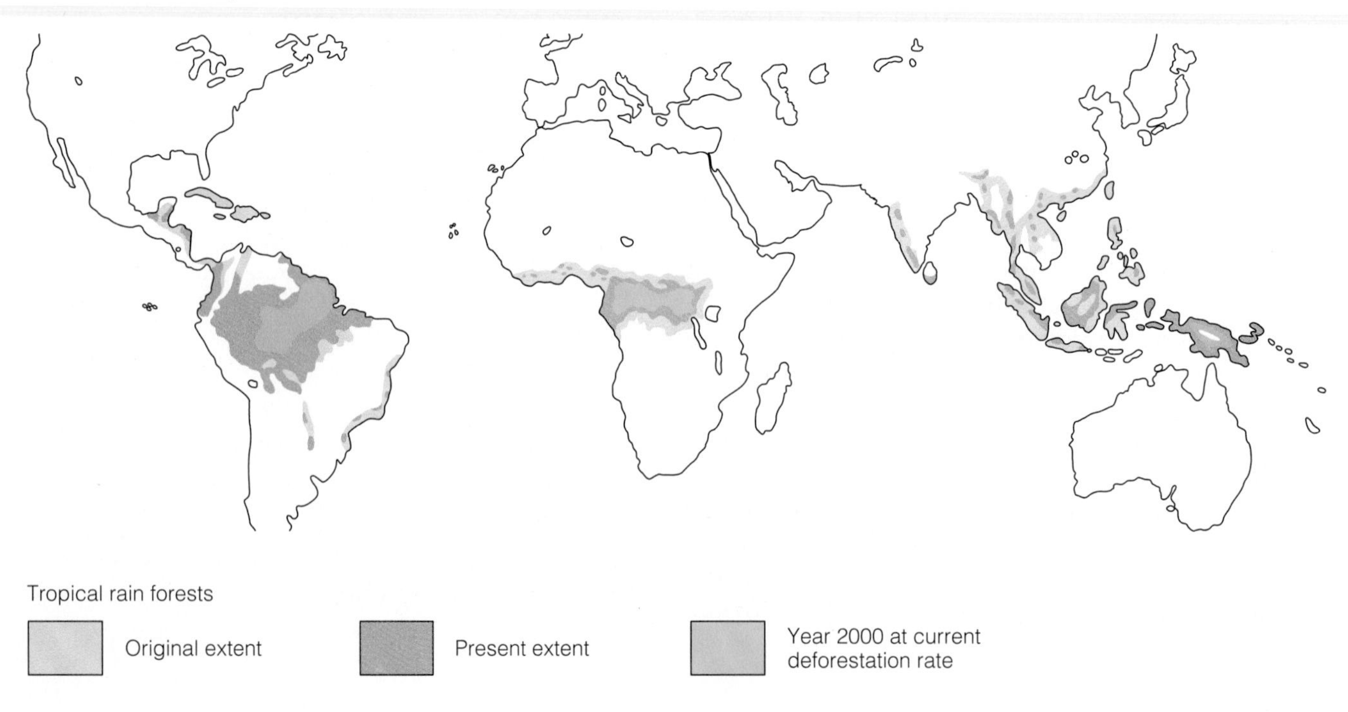

FIGURE 17-4 Soil Erosion Millions of acres of farmland are destroyed each year because of poor land management practices that lead to severe soil erosion.

and rangeland are lost every day. That's equivalent to 1.25 million acres a year, or a strip 0.3 mile wide extending from New York City to San Francisco.

Although these trends bode poorly for us, the loss of topsoil and farmland can be stopped, and soils can be replenished. However, worldwide conservation efforts are needed as are population growth measures, which help to reduce the conversion of farmland to other uses such as housing. Additional solutions shown in Table 17-1 could alleviate world hunger and could also reduce the loss of productive farmland.

Water. Many areas of the world are facing water shortages or will soon face them as population increases. Water shortages generally result from too many people drawing on limited water supplies—either groundwater or surface waters, such as lakes and rivers. Water shortages affect cities and towns, businesses, and farming operations and can have severe economic impacts. Fortunately, there's much that we can do to prevent shortages, for example, using water much more efficiently, especially on farms, one of the world's largest users of water. Measures to conserve water around our homes, especially outside water, can help as well.

Oil. Oil is the lifeblood of modern society. In the United States, for example, oil supplies approximately 40% of our annual energy demand. But oil supplies are finite. No one knows exactly how much oil is left in the Earth's crust, but many analysts believe that world oil supply will peak between 2005 and 2010. After this time, demand will exceed supply, causing prices to increase. So what do we do?

The first step in meeting future demand is to make current energy use much more efficient. The efficient use of oil and oil products, such as gasoline, diesel fuel, and home heating oil, can help us stretch current oil supplies considerably. More efficient cars could cut demand enormously.

Another important step in increasing energy efficiency is mass transit buses and trains to transport commuters to and from work. Mass transit is four times more efficient than the automobile.

Alternative fuels will also help meet future demand for transportation fuel. One example is ethanol. Produced from corn, wheat, and other crops, ethanol can power cars and trucks. Grown on farms, ethanol could virtually replace gasoline.

Another potentially important fuel is hydrogen. Made from water, hydrogen is the ultimate in renewable fuels. It burns cleanly and is available from an abudnant resource.

Renewable energy such as electricity made from sunlight could provide enormous amounts of fuel in the years to come. According to one estimate, nonrenewable energy reserves (coal, oil, natural gas, and so on) would provide the equivalent of 8.8 trillion barrels of oil. Renewable energy could provide 10 times that amount of energy every year!

FIGURE 17-5 Farmland Conversion Cities are often surrounded by excellent farmland soils that drain well and are flat and highly productive. Many of the features that make soils suitable for farmland also make them suitable for building. This scene, unfortunately, is all too common in expanding urban areas.

Pollution of Our Environment

Several forms of pollution are overwhelming nutrient cycles, poisoning other species and ourselves. One of the most serious is the greenhouse gases responsible for a phenomenon called *global warming*.

Global Warming. Greenhouse gases are natural and human-made pollutants that trap heat in the Earth's atmosphere, causing it to warm. They act much like the glass in a greenhouse, hence their name. One of the most significant greenhouse gases is carbon dioxide, a pollutant that arises from the combustion of fossil fuels, such as gasoline.

Since the 1880s, scientists have known that carbon dioxide is a greenhouse gas. Today, we know that a little carbon dioxide in the atmosphere is essential to life on Earth. Without this gas, the planet would be about 55°F (30°C) cooler than it is. Too much carbon dioxide, however, leads to overheating, called **global warming**. Over the past 100 years, global carbon dioxide levels have increased over 32%, as a result of the combustion of fossil fuels and deforestation (FIGURE 17-6A).

Several other pollutants also contribute to global warming. Methane, for example, is released from livestock and the manure they produce. Humankind's nearly 1 billion cattle annually release huge amounts (about 73 million metric tons). Methane is 20 times more effective at trapping heat than carbon dioxide.

Chlorofluorocarbons, or CFCs, are also greenhouse gases. These chemicals have been used for many years in refrigerators, air conditioners, and freezers. One class of CFCs was even used to clean circuit boards for computers and other electronic equipment. Best known for their effect on the ozone layer, discussed shortly, CFCs also trap heat in the atmosphere, contributing to global warming.

Most atmospheric scientists believe that the dramatic increase in the release of greenhouse gases is responsible for a rise in global temperature (FIGURE 17-6B). Continued release could result in further increases with potentially devastating effects.

According to the latest projections based on current increases in greenhouse gases, average global temperatures could be 2.7 to nearly 11°F hotter by 2100. As the planet warms, some scientists predict that global rainfall patterns will shift, making some areas wetter and others hotter and drier than others (FIGURE 17-7). Agricultural areas that receive too much rainfall or too little could experience dramatic decreases in output, reducing food supplies worldwide.

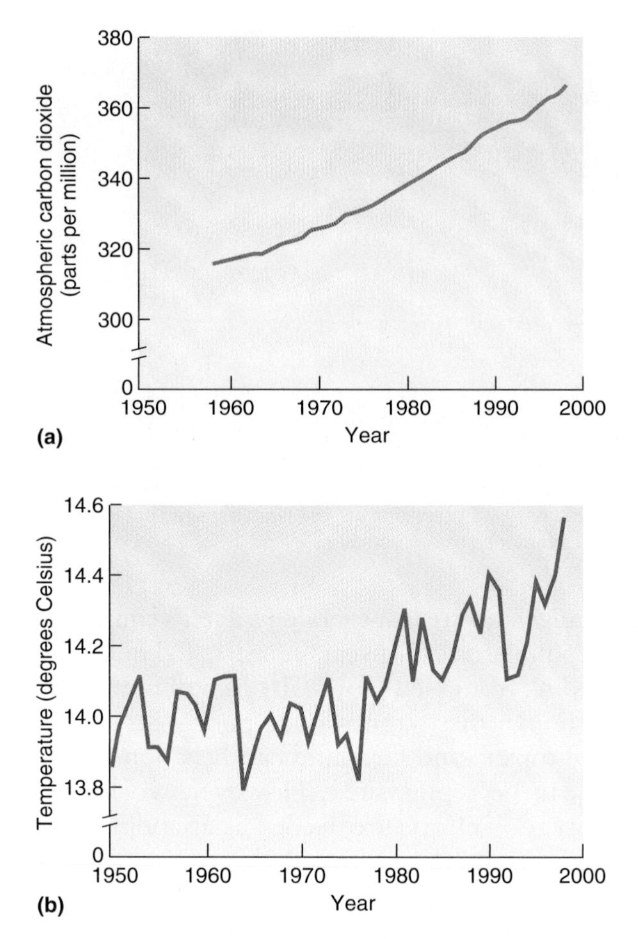

FIGURE 17-6 Global Carbon Dioxide and Temperature Trends *(a)* Carbon dioxide levels in the atmosphere have risen dramatically since 1958. *(b)* Graph of average global temperature since 1950.

Rising global temperature could also cause further melting of glaciers and the polar ice caps (FIGURE 17-8). Melting ice may result in an increase in sea level. By 2100, scientists predict that the sea level will rise 50 centimeters (2 feet) because of global warming.

In the United States, approximately half of the population lives within 50 miles of the ocean, and many large cities such as Miami are located only a few feet above sea level (FIGURE 17-9). Rising seas could flood these and other low-lying regions. Expensive dikes and levees would be needed to protect cities such as Miami. Other coastal cities would have to be rebuilt on higher ground, at a staggering cost. A rising sea level would be particularly hard on Bangladesh and other Asian countries with extensive lowland rice paddies.

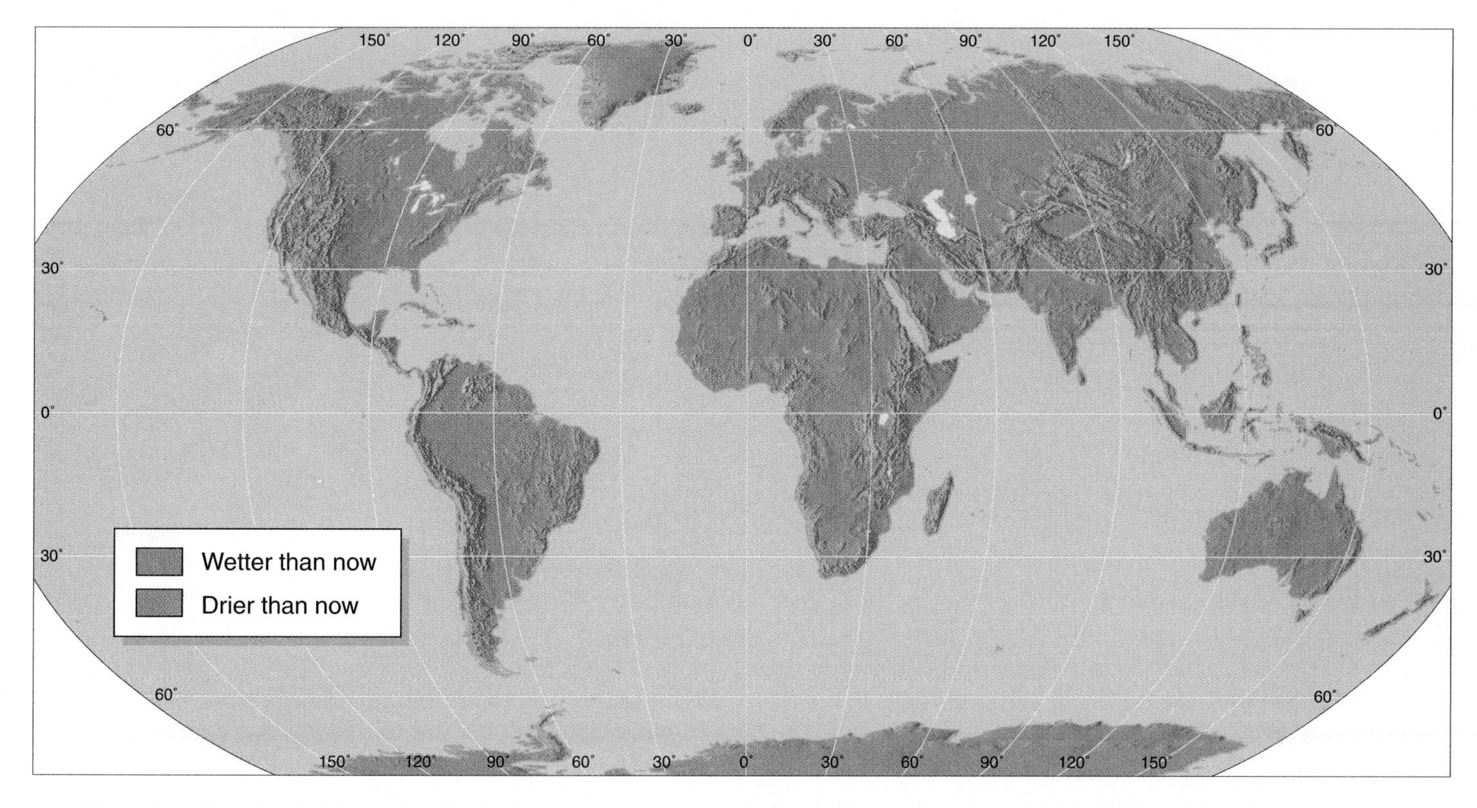

FIGURE 17-7 Changing Rainfall Patterns Possible future changes in climate resulting from the greenhouse effect. This map is based on climatic conditions thought to have existed 4500 to 8000 years ago, when the average temperature was highest. Note that this is a generalized map and may miss regional changes.

FIGURE 17-8 Mount Rainier Glaciers This enormous glacier and others like it could melt as the Earth gets warmer, raising the sea level.

FIGURE 17-9 Miami Underwater? A rising sea level due to global warming could flood many coastal cities the world over.

Rising temperatures would also have an impact on many species of plants and animals. Although some species might adapt or find new habitat, many plants could be wiped out, as temperatures would probably increase faster than they could adapt. Forests could dry out, die off, and burn.

Some pest species may actually thrive amid global warming, for example, organisms such as mosquitos that are responsible for carrying diseases like the West Nile Virus. Some diseases now restricted to tropical areas could invade new territory.

Global warming could cause more dangerous heat waves, which could kill hundreds, even thousands of people worldwide. Approximately 700 people die from summer heat waves in the United States each year.

Global warming may also trigger more violent weather, especially hurricanes and tornados that cause billions of dollars worth of damage.

These predictions, many of which are already becoming apparent, sound dire. A small number of scientists believe that we don't know enough about the atmosphere and the impact of greenhouse gases to know for certain whether the changes that are taking place are due to greenhouse gases and deforestation. They also note that global climate is extremely complex and several factors could dampen the warming trend. Other factors could worsen it.

Slowing down, even halting global warming requires numerous measures. We can, for example, plant more trees and use energy more efficiently to reduce atmospheric carbon dioxide. Alternative sources of fuel, such as hydrogen, solar energy, wind, and ethanol, could help.

Acid Deposition. In many parts of the world, fishes have vanished from lakes that have turned acidic (**FIGURE 17-10**). The acids come from the skies where they are formed from two air pollutants, sulfur dioxide and nitrogen dioxide. These pollutants arise chiefly from the combustion of fossil fuels: coal, oil, and natural gas in power plants, factories, cars, aircrafts, and homes.

In the atmosphere, these pollutants react with water and oxygen to form sulfuric and nitric acids. Rain and snow wash these acids from the sky. Fog and clouds also carry heavy loads of acids that are deposited on trees and buildings.

FIGURE 17-10 Victims of Modern Society These fish were killed by acids from acid deposition.

In the United States and Europe, acid rain, snow, and clouds have been worsening for five decades as a result of increased fossil-fuel combustion. During this time, the regions affected by acid rain and snow have grown larger. In addition, the strength of the acids has increased (**FIGURE 17-11**). Today, acid deposition is commonly encountered downwind from virtually all major population centers.

Acid deposition kills fish and other aquatic organisms. Acids on land dissolve toxic minerals such as aluminum from the soil and wash them into surface waters. Aluminum causes the gills of fishes to clog with mucus, suffocating them. Extensive acidification of lakes and rivers is putting resort owners out of business in the Northeast, upper Midwest, and southern Canada.

Acids damages buildings, statues, lakes, trees, and crops and may be responsible for the massive forest diebacks occurring throughout the world (**FIGURE 17-12**).

Reducing the deposition of acids will require a dramatic reduction in the release of sulfur dioxide and nitrogen dioxide from power plants, factories, and automobiles. Sulfur dioxide release can be achieved by installing pollution control devices on power plants or using low-sulfur coal, which many utility companies now do. Using energy more efficiently helps curb pollutants, as do efforts to tap into renewable fuels such as wind power and solar energy.

Tío Fernando is my uncle, my *mamá's* brother.
Abuela is their mother.
She raised them on *la isla*.

"¡Bienvenidas!" Tío Fernando welcomes us.
He and my *mamá* grew up in this house
with Abuela and Abuelo, my grandfather.
Abuelo died before I was born.
Now *tío* Fernando lives here with his family.
I think he looks like my *mamá,*
except he has a beard.
"El osito," Abuela calls him — the little bear.

ESCUELA PRIMARIA UNIDAD

Abuela shows me all around.
In the front room, she and Abuelo
used to run a little store.
On the wall, next to a picture of the store,
is a painting of *tío* Fernando with a giant fish.
"¡Qué pescado!" Abuela says,
telling me what a fish it was.
Tío Fernando found it in a shallow stream.
He brought it home to keep for a pet.
Abuela said the fish would be
happier in the river.
Tío Fernando was sad to see it go,
so Abuela painted the picture for him.

"Los niños," Abuela says,
showing me a picture of some children.
It's my *mamá* and *tío* Fernando
playing in a fountain.
Abuela and Abuelo built the fountain
with stones from the rain forest.
It is still in the yard.
"Es mágica," Abuela says.
The fountain does seem magical.
The water splashing over stones
sounds like birds singing.
Now Abuela wants to show me more of *la isla.*
Elena says she'll meet us later at the beach.
"¡Que disfruten!" she calls. She wants us to have fun.